D0086762

ROHM AND HAAS

A CENTURY OF INNOVATION

R & H
OHM
AND
AAS
CO. INC

ROHM AND HAAS

A CENTURY OF INNOVATION

REGINA LEE BLASZCZYK

FENWICK

Rohm and Haas Company
100 Independence Mall West
Philadelphia, Pennsylvania 19106

First edition
Printed in China

This book was printed on Novatech Matt Art paper using soy-based inks. Novatech is manufactured using Elemental Chlorine Free (ECF) pulp sourced from sustainable, well-managed forests. Soy ink is an alternative to petroleum-based inks that is derived from a renewable source, is low in volatile organic compounds (VOCs), and is more easily recycled.

18 17 16 15 14 13 12 11 10 09 1 2 3 4 5

ISBN: 978-0-9749510-8-9
The Library of Congress Cataloging-in-Publication data is on file.

Fenwick Publishing Group, Inc.
3147 Point White Drive, Suite 100
Bainbridge Island, Washington 98110

Fenwick produces, publishes, and markets custom publications for corporations, nonprofit organizations, family foundations and individuals.

www.fenwickpublishing.com

President and Publisher: Timothy J. Connolly
Vice President, Development: Sarah Morgans
Production Manager: Patrick R. Duff
Designer: Kevin Berger
Proofreader: Lisa Wolff
Indexer: Ken DellaPenta

Fenwick Publishing extends grateful appreciation to Carmen Ferrigno, Anthony Sergi, Shana Pote, and Chad Stuckey for their efforts and insights during the production of the book.

For assistance in research and writing, Regina Lee Blaszczyk extends her thanks to Arnold Thackray, Gwen Burda, Phyllis Dillon, Marthenia Perrin, Nicholas Schlosser, Kathryn Steen, and Patricia Wieland. She also thanks Professor Patrick Fridenson of the École des Hautes Études en Sciences Sociales in Paris, Jon Williams at the Hagley Museum and Library, Linda Skolarus at the Henry Ford Museum, Jim Roan at the Smithsonian Institution Libraries, and design historian Phyllis Ross for sharing sources and expertise.

ALTHOUGH CONSUMERS NEVER SEE ROHM AND HAAS'S NAME ON THE LABEL, THE COMPANY DEVELOPS AND MANUFACTURES SPECIALTY MATERIALS THAT IMPROVE COUNTLESS PERSONAL CARE PRODUCTS AS WELL AS PRODUCTS OFFERED BY THE BUILDING AND CONSTRUCTION, ELECTRONICS, FOOD, INDUSTRIAL, PACKAGING AND PAPER, PHARMACEUTICAL AND MEDICAL, TRANSPORTATION, AND WATER INDUSTRIES. HERE, SENIOR SCIENTIST MIAO WANG LABELS A HAIRSPRAY BOTTLE FOR A FORMULATION COMPATIBILITY TEST.

Fisher
16B

RESIN RESEARCH LABORATORY
222 WEST WASHINGTON SQ.
PHILA. PA.
No. 3367

TABLE OF CONTENTS

ROHM AND HAAS COMPANY | **FOREWORD**

This year we mark the 100th anniversary of Rohm and Haas. Ours is a rich history begun with a one-on-one collaboration between a chemist named Otto Röhm and a businessman named Otto Haas to meet a critical business need through chemistry. What these two men started has grown in the ensuing 100 years into a company and a philosophy that has improved the lives of people the world over.

From that initial bond, the two entrepreneurs cracked a code and created a formula for success that remains the hallmark of Rohm and Haas's business today: use close relationships with markets and customers to develop singularly unique, technically advanced materials that solve seemingly unsolvable problems. And what started as a smart business practice has become a defining company culture.

Today, Rohm and Haas scientists, engineers, operators, and marketers live and work in every corner of the world. Their knowledge and expertise are shared freely across the company and fuel each successive team of inventors and problem solvers. As I look back on the accomplishments of more than five generations of employees, I am proud to see that we have helped shape the history and the lives of many millions of people. From the communities in which we live to the markets we serve and the consumer products we help create, our legacy is tangible and our accomplishments profound.

Any history, no matter how rich and expansive, ultimately guides and informs the present. Throughout this book, which is replete with images that capture our history during the last century, we also have placed photos of Rohm and Haas employees today. Together, these past and present images capture a culture that has thrived on innovation, collaboration, and a fierce dedication to improving the world in which we live.

We hope you recognize that today's employees stand upon the shoulders of those who came before and the innovations they left as their legacy. For those of you who are here in 2009, we want you to know that your work has set the stage for a new era of innovation dedicated to tackling the most important issues of our times. Your accomplishments mean more than success for Rohm and Haas in its many markets and regions; they mean your contributions have played a principal role in improving the lives of millions of people in ways few may see but all experience.

Raj Gupta
February 1, 2009

During the span of a century, Rohm and Haas has grown from providing a single product used in the tanning of leather into an array of specialty materials that make life better for people around the world. Opposite: A worker at Rohm and Haas's Marlborough facility prepares to enter a clean room.

ROHM AND HAAS COMPANY | **PREFACE**

This book tells a wonderful story that needs to be widely known. Advances in living standards that the world has enjoyed in the last one hundred years—and advances that beckon us in the next one hundred—owe a good deal to the chemical industries. Within those industries it is the specialty chemical companies that have held pride of place. And within those specialty chemical companies no name is more distinguished than Rohm and Haas, which celebrates its centennial year in 2009.

The story of Rohm and Haas is a very human story that begins in 1909 with a young European immigrant to America and ends with a thriving, diversified, global corporation led by another immigrant, this time from Asia. This is a story of entrepreneurship and perseverance, of scientific research and technological ingenuity, of risk and reward, and of vision and global leadership. Above all, it is the story of innovation through collaboration between the chemical manufacturer and the customer, the story of a firm that has a long tradition of developing specialty products that exactly meet its customers' needs. F. Otto Haas, the founder's elder son, said it well: "The key to the whole company is really studying an industry . . . finding what its needs are and then developing a chemical to meet those needs. . . . That's exactly what we do."

Telling this story was a privilege and a pleasure. Working closely with Brian McPeak, vice president for communications, I was blessed with unrestricted access to Rohm and Haas's corporate archives and the opportunity to interview many top executives from throughout the company's history. In boxes of yellowing old documents I experienced the thrill of the hunt as I explored rich veins of evidence that speak to the company's long-standing commitment to customer service.

Most important, I was delighted that Rohm and Haas gave me the freedom to develop the story that emerged from the evidence in my own way. The real history of Rohm and Haas is more than a simple narrative of events: it is a tribute to creative people committed to the needs of their customers. The material that follows was designed to meet the exacting standards of the historical profession, while engaging readers with an exciting story about one of the chemical industry's leading firms.

Enjoy!

Regina Lee Blaszczyk
January 1, 2009

During the past century, Rohm and Haas has been not only the developer and manufacturer of cutting-edge products, but also the beneficiary of the technology. Here, a Rohm and Haas scientist swings a Plexiglas safety shield out of the way in order to adjust a testing apparatus, circa 1958. In addition to serving as an important material used in World War II military aircraft, Plexiglas shatter-resistant acrylic plastic, developed by Rohm and Haas, was used throughout the company's labs to protect against flying glass shards and spattering liquids.

01 LAUNCHING IN LEATHER

Otto Haas, Entrepreneur

Otto Haas was a born salesman. Just as a horse trader could size up a stallion, Haas knew how to read a customer. The man standing before him, Arthur C. Best, the superintendent at the Robert H. Foerderer tannery in Philadelphia, was a tough prospect to read. In his thick German accent Haas explained the advantages of Oropon, a new synthetic bate made by his partner, Otto Röhm, in Darmstadt, Germany. The new bate was an effective, sanitary alternative to the dog manure and bird feces that tanneries had used to soften hides and skins since time immemorial. But on October 24, 1910, Best was not in the mood for a sales pitch. Intuition told Haas to take it easy, lest his potential customer slip away. Haas had previously sold small quantities of Oropon to the Foerderer tannery, which experimented on some hides with it. Then things went cold. The foreman clammed up,

OTTO HAAS, SHOWN IN THE 1920S SURROUNDED BY WORKERS AND THEIR FAMILIES AT HIS PLANT IN BRISTOL, PENNSYLVANIA, BUILT ROHM AND HAAS ON A SINGULAR PRODUCT, OROPON. OROPON OFFERED A SANITARY ALTERNATIVE TO THE NOXIOUS SUBSTANCES TANNERIES HAD USED TO SOFTEN HIDES AND SKINS IN THE LEATHER-MAKING PROCESS. PREVIOUS SPREAD: THE PROCESSING OF HIDES IN AN AMERICAN TANNERY INCLUDED THIS STAGE IN THE BEAM HOUSE. WORKERS LAID HIDES OVER "BEAMS" AND SCRAPED THEM WITH BLUNT KNIVES TO REMOVE THE FLESH AND HAIR. THE HIDES WERE THEN TREATED WITH OROPON TO MAKE THEM RECEPTIVE TO THE TANNING AGENTS.

Otto Röhm, shown right in a studio portrait taken in Germany in 1908, formed a transatlantic partnership with Otto Haas in 1909. Röhm worked on the chemistry of their synthetic tanning bate from Germany while Haas sought customers in Philadelphia, the global center for the production of glazed kid. Philadelphia tanner Robert H. Foerderer built the market for kid leather with extensive consumer advertising. The colorful booklet below, published circa 1897, told consumers how to care for stylish footwear made from his Vici Kid brand leather.

and Haas's entreaties to top managers H. W. Trudell and Perceval Foerderer fell on deaf ears. From spring to early fall Haas doggedly trekked up to the Frankford section of Philadelphia to call on the company, always returning to his office at 202 North Second Street without the order. This time was no different. Best did not "want to make any changes at present" because his hands were "full with other troubles."[1]

Haas remained calm, resolved to persevere. Philadelphia was a global center for the production of glazed kid—a soft, pliable goatskin for fashionable ladies' shoes and gloves. Every year the city's tanneries turned sixty million goatskins into kid leather for consumption at home and abroad. As a contemporary observer noted, "More goatskins are consumed in Philadelphia than in any other city in the world." The Foerderer tannery was the largest in the city, tanning no less than three hundred thousand skins every week.[2]

In the late 1880s the tannery's founder Robert H. Foerderer—a German immigrant like Haas—helped bring the light leather industry into the modern era. He built a high-tech facility to process goat- and sheepskins with chrome tanning, an innovative mineral-based technique that produced better leather than the standard vegetable tannin. Foerderer, borrowing from Latin, named his brand Vici Kid, or "I conquered kid." Before long the leather industry adopted the word *vici* as a synonym for chrome-tanned kid. When Robert died in 1903 at age forty-three, his son Perceval inherited a leather dynasty and a world-famous brand.[3] Haas determined that if Oropon were to be used on Vici Kid, other tanners might try it. As fall turned into winter, Haas waited, eyes fixed on this prize customer.

German Roots

Otto Haas first came to the United States in 1901 at the age of twenty-nine. The son of a German civil servant, Haas had started out as a clerk in Stuttgart's Hofbank, handling English accounts and mastering the language. He took pride in his ability to absorb large quantities of information and process it faster than his peers. Ambitious, determined, and hard driving, he decided to scout out the fabled opportunities of the new world. Initially he worked as a clerk for G. Siegle and Company, a dye and chemical manufacturer on New York's Staten Island. Two years later he moved to the Manhattan office of Sulzberger and Company, a large German-American meatpacker in the export trade.[4] In these jobs Haas perfected his knowledge of American business practices, worked on his English, and absorbed the social and cultural surroundings.

Back in Germany, friend and future business partner Otto Röhm followed a different path. Röhm was a Ph.D. chemist and a gifted scientist who had written his 1901 thesis on acrylic acid, thereby laying the groundwork for his later innovations in plastics. First meeting in 1900 through Röhm's cousins, the Kohler family, the two Ottos quickly became staunch friends.[5] In due course the drive and business acumen of Haas were to make a natural foil to the scientific and technical talent of Röhm.

Leather was everywhere one looked at the start of the twentieth century. However, leather tanning was a notoriously dirty activity, creating noxious waste and foul smells. By the early 1900s the leather industry had several branches. Heavy-leather makers processed unwieldy cattle skins, or hides, into sturdy industrial belting, horse tackle, heavy luggage, and the soles of footwear. Specialists in calfskins, goatskins, sheepskins, and the more exotic ostrich skin turned these lighter materials, or skins, into fancy leathers for shoe uppers, shoe linings, gloves, upholstery, smaller luggage, billfolds, and change purses.[6] The two branches were distinct and independent, as noted in 1900 by a U.S. census report: "A sole leather tanner can no more make glove leather than a blacksmith can make watches."[7]

Leather making consisted of three main steps: beaming, tanning, and finishing. Each laborious task was done by hand, as a hide or skin traveled slowly through the tannery. Workers in the "beam house" prepared the hides and skins, which arrived in port dried or salted. They were soaked, washed, softened, and dehaired with lime and arsenic, rendering them open and vulnerable. The foulest part of beam-house work was the bating process. The workers bated, or drenched, the animal skin in a liquid mixture that softened the pelt and opened the pores for the tanning agents. Then in the tanning rooms hides and skins were turned into leather by waterproofing, coloring, resoftening, and glazing. Since ancient times tanners had used dog manure, hen feces, or pigeon droppings as the main ingredient in the bating slurry, without understanding why it worked. Cheap and effective, the excrement, imported from Turkey, was also highly unsanitary and could even expose workers to anthrax. If hot weather arrived unexpectedly, the manure bate fermented and damaged the skins irrevocably.[8]

Workers test Oropon in a tannery at Esslingen, near Stuttgart, above. Two men bate skins in a wooden tub while a third dehairs a skin over a beam, circa 1907.

The peculiarities of leather chemistry were to capture Röhm's imagination. Initially restless, the young scientist took a series of jobs—a brief stint at Merck and Company in Darmstadt, a time teaching high-school science, and a spell in a government agricultural lab—before moving to Gaisburg in 1904 to work as an analytical chemist for the Municipal Gas Works. In his free time Röhm's curiosity led him to the nearby tanneries, where he noticed an odor similar to that of the "gas water" of his day job. From a tanning chemist Röhm learned about the primitive nature of the bating process and speculated that he could create a more effective, sanitary bate. Soon Röhm was burning the midnight oil, hunched over his own experiments in the gasworks lab. He focused on ammonium sulfate and its role in the malodorous "gas water" of coal-gas production. His experiments attracted the notice of several tanneries, including the glove-leather works of H. Bodner. Impressed by the young chemist's ideas, Bodner's owners invited Röhm to continue his work at their Esslingen tannery, just outside Stuttgart.[9]

By May 1906 Röhm's experiments at Esslingen were so promising that he filed for a German patent on his bating process, based on gas water. Later that month

Bodner acquired the exclusive rights to sell Röhm's bating agent in Germany and Luxembourg. Quickly Röhm conveyed the news in an upbeat letter to Otto Haas in New York. "According to the unanimous opinion of the Gaisburg and Esslingen tanners," he wrote, "my process does not only represent a fully adequate substitute for the dog manure but surpasses that action materially." In accord with plans they had mutually agreed on, Röhm then urged that Haas return to Germany and that the two friends form a fifty-fifty partnership to make this new miracle bate. Röhm also laid out an American marketing plan for Oroh, so called by combining his and Haas's initials, O. R. and O. H.[10]

Alas, youthful exuberance had led Röhm to jump the gun. After a few months in storage seven thousand skins that had been bated with Oroh failed to hold their dye, just as Haas arrived back in Germany in January 1907. The two men labored side by side in the lab to resolve Oroh's problems. "We are working closely together," Haas confided to Josen Hans, an old mentor from his teen years as a pharmacy assistant. "At times it looks like the American way. We work regularly [and] arrive too late for our midday meal." The long hours and exhaustive tests produced nothing but frustration. By March 1907 Haas became "somewhat nervous and impatient," while Röhm continued to plug away. In May, Röhm hinted at the first advance in months: "If the skins and hides are first placed in a dung bate and then into my product, the long desired effect is achieved. Yesterday's experiment indicates that these skins take up something from the manure which then acts together with my bate. In what manner would I find this material?"[11]

Röhm suspected that enzymes in the manure were the active agent. He knew about the work of Professor Eduard Büchner at the University of Tübingen, whose research on the enzymes in yeast would later earn him a Nobel Prize. Inspired by Büchner's work, Röhm persevered. Another notebook entry in late May—

"pancreas gland acts like manure"—documents a eureka moment. Röhm obtained some pancreases from an animal slaughterhouse, whose workers were accustomed to taking them home to eat as sweet meats. He ran the sweet meats through a grinder, squeezed out the juice, and mixed the resulting liquid with mineral salts. Röhm patented the new process and trademarked the product as Oropon, which combined his initials with the Greek word for juice, *opas*.[12]

With invention in hand Otto Röhm and Otto Haas signed a partnership agreement on August 30, 1907. They decided to make Oropon in a small house they rented for the purpose in Esslingen and to market their bate to German tanneries.[13] For this latter task Haas drew on his knowledge of both German dyestuffs companies and Siegle's American office to initiate the hands-on sales techniques that would distinguish Rohm and Haas as a specialty chemical company for the next hundred years. (The company name initially was Röhm and Haas, with the

Otto Röhm and Otto Haas not only developed Oropon for use in tanneries like the Rodel Leather Manufacturing Company, opposite; they also discovered new uses for it. When tannery workers began using Oropon to treat minor skin injuries, Röhm was inspired to create the topical ointment Pancrazym, left, from pancreatic enzymes. Above: Röhm is shown at left in the doorway, and Haas is in the window at right at their first office and factory in Esslingen, Germany, circa 1907.

ROHM AND HAAS COMPANY | **TODAY**

MANUFACTURING STRENGTH

Among the materials that Rohm and Haas has developed for customers are technically advanced industrial coatings that can withstand harsh conditions and adhere to challenging substrates. Here, coatings scientist John Halligan makes a polymer for architectural wall coatings.

The secret was to spend considerable time in the customer's plant with the chemist, foreman, and workers. This way Otto Haas learned the customer's realities, saw how the proposed science-based solution performed under distinctive factory conditions, and reported back to his colleague in the laboratory. The product's formula could then be adjusted until it fully met the customer's needs.

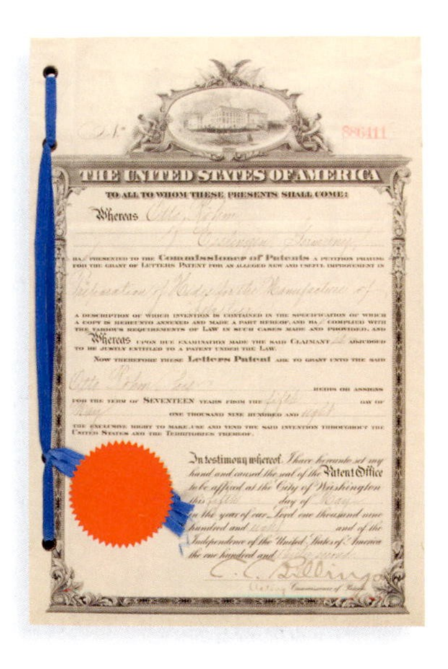
THE UNITED STATES OF AMERICA
TO ALL TO WHOM THESE PRESENTS SHALL COME:

The United States Patent Office awarded Rohm and Haas its first U.S. patent on May 5, 1908. U.S. Patent 886,411, above, was granted to Otto Röhm for the "Preparation of Hides for the Manufacture of Leather."

umlaut, but in later years the umlaut was dropped.) The secret was to spend considerable time in the customer's plant with the chemist, foreman, and workers. This way Otto Haas learned the customer's realities, saw how the proposed science-based solution performed under distinctive factory conditions, and reported back to his colleague in the laboratory. The product's formula could then be adjusted until it fully met the customer's needs.

Haas developed the principle of sales plus service as he plunged into the leather business. He traveled to numerous German tanneries, explaining why they should buy Oropon, a more expensive option than manure. He showed the men in the beam houses how the synthetic material worked. Mired in tradition, tanners distrusted newcomers and were a hard sell. Other times Haas bumped up against Mother Nature and Father Time. Once he trudged two miles in the snow with a fifty-pound glass carboy filled with liquid Oropon on his shoulder after missing the last stagecoach to the tannery.[14]

Building the market for their product was difficult. On several occasions the partners almost gave up, but slowly enough tanneries invested in Oropon to make it a commercial proposition. Kid tanners provided the largest market. The product's enzymatic action was ideal for achieving the ultrasoft effect desired in kid gloves. By listening to the customer Haas realized that different tanneries needed different grades of Oropon. Back in Esslingen, Röhm modified the master formula and created such variations as Oropon B or D to suit specific production requirements. With the German market secured, Haas began selling Oropon in England in 1908, and Röhm's brother, Adolph, started a sales office in Lyon, France, in 1909.[15] That year Rohm and Haas opened a larger factory in Darmstadt, an industrial city in the Ruhr Valley, near several tanneries. The new facility had a miniature tannery, allowing Röhm to experiment as needed. The partners' division of labor was established. Röhm oversaw research and production, while Haas focused on building a market. And most important of all he could return to America confident that Rohm and Haas had a product that leather customers wanted.

Next Stop, USA

When the Rohm and Haas sales office opened in Philadelphia on September 1, 1909, the United States and Germany were on parallel economic paths. Both countries were becoming major industrial powers, challenging Great Britain's place as the world's leading manufacturing and consumer society. The Civil War in the United States and the Prussian War in Europe had jump-started industrialization. The two nations also shared social networks and cultural traditions. Millions of Germans had moved to North America, from colonial times onward. German businessmen relished America's entrepreneurial culture and established successful companies: breweries in Milwaukee and St. Louis and tanneries along the East Coast.[16] Germans figured out how to mix Old World discipline, attention to detail, and superb craftsmanship with the fast pace of life and the unbridled American rush into the future.

Germany dominated the high-tech industry of the day: chemicals. After Bismarck united the German states

Otto Haas commissioned Gerrit A. Beneker, the foremost painter of American industry during the 1920s, to paint a series of pictures showing Bridesburg employees. Completed in 1930, his painting of George Trapp, above, captures the integrity of the blue-collar worker. Dressed in work clothes, Trapp stands with the tools of his trade, the green glass carboys and wooden crates that were used to ship oil of vitriol.

The Boss at Bridesburg

By the time Otto Haas settled in the Quaker City, the Delaware River village of Bridesburg had become a thriving industrial town. Its main employer was an impressive chemical plant, begun by Charles and Frederick Lennig in 1842 to make basic chemicals for industry—products like soda ash, alum, and sulfuric acid, commonly known as oil of vitriol. By the early twentieth century the sprawling Lennig works was testimony to Philadelphia's role as "the workshop of the world."

Bridesburg was home to German and Polish immigrants, who worked at the Lennig facility, the docks at nearby Port Richmond, and the Foerderer tannery a short trolley ride away in Frankford. When Haas purchased Charles Lennig and Company in 1920, he felt at home among workers who had come from the German chemical industry. In due course he bought land adjacent to the aging Lennig works and created an enlarged Rohm and Haas. Today "Bridesburg" is familiar shorthand for the chemical manufacturing that forms a core competence at the heart of Rohm and Haas.

Over the years Otto Haas himself became a familiar figure in Bridesburg, as he undertook meticulous hands-on inspections of operations and personnel. He never learned to drive, but every week he was chauffeured to the plant in the company car to meet with the manager and inspect the facilities.

Employees were on pins and needles during these visits. Haas—with dry humor—would alternate between the roles of the stern paymaster and the benevolent paternalist. An employee "underground" kept track of his movements, phoning from department to department about his imminent arrival. Ashtrays disappeared and rooms were ventilated because everyone knew Haas disapproved of smoking. One time a young clerk hastily snuffed out a cigarette and shoved it in his desk. As he talked with the boss, smoke started curling out of the drawer. Haas let the conversation drag on, watching the nervous clerk with amusement. Finally he said, "Young man, I think that you had better call the fire department. Your desk is on fire!"

Otto Haas's old-world civility and lifelong engagement with his enterprise are well caught in the memories of Eugene Kasperowicz, who started at Bridesburg in 1947. "When I was leaving my work area for home, I'd cross paths with an older gent walking to the production area. He was dressed in a button-up dark gray business suit, hat, high-button collar shirt, and tie. As we approached one another, we would make eye contact. I wished the man 'good night,' and he would curtly nod and crisply reply, 'Good night sir.' I thought that it was nice of him to address me, a kid, as sir. Sometime later, my shift foreman informed me that I had been trading good nights with Mr. Haas."

in 1871, the national government fostered industrial growth by establishing protective tariffs and turning a blind eye toward cartels. German universities cooperated with industry, sharing researchers, ideas, laboratories, and patents, thereby making big advances in synthetic organic chemistry. Such firms as Bayer, BASF, and Hoechst controlled the global market in synthetic dyes, intermediates, and pharmaceuticals. In leather, however, American ingenuity trumped German expertise. Using innovations like chrome tanning, mechanization, and the division of labor, U.S. manufacturers developed a comparative advantage. American tanneries were larger, more efficient, and more creative than their European counterparts.

The leather industry was concentrated in production centers in the Midwest and the Northeast. By 1900 the Chicago-Milwaukee tanneries were owned or allied with the giant meatpackers. These tanneries produced cattle-hide leather, using raw materials from their own packinghouses. Eastern tanneries developed local specialties: sole leather around Boston; patent leather in Newark, New Jersey; fine glove leather in Gloversville, New York; and kid leather in the Delaware Valley. The kid sector was particularly creative and adaptive. Importers landed goat, sheep, calf, and ostrich skins—from southeastern Europe, the Middle East, Latin America, Africa, and Australia. These exotic materials were then turned into fine-grained leather at locations in the Delaware Valley, such as Wilmington, Camden, and Philadelphia.[17]

The area's kid factories operated at a favorable moment in American economic history. Deflation allowed Americans to buy more, and the birth of mass retailing put many products within reach of ordinary people. New merchandisers—mail-order catalogs, chain stores, and department stores—took advantage of the favorable conditions. Retailing entrepreneurs like Richard Sears, Frank W. Woolworth, and John Wanamaker revamped distribution and reduced retail prices. In 1909 the nation's largest mail-order house, Sears, Roebuck and Company, offered ladies' oxford shoes in vici kid for a mere dollar and a half.[18]

American kid-leather factories embraced new technologies because their survival depended on novelty. To be fashionable, shoemakers designed women's pumps, slippers, and boots with many types of leather. Anything a tannery could generate in the way of fresh styles, grains, and colors at a lower price was welcome. By 1917 the United States made 80 percent of the world's supply of these fine-grained leathers, with the Philadelphia-area tanneries accounting for 60 percent of that output.[19]

American trade journals like the Shoe and Leather Reporter *were filled with sumptuous advertisements, like the one below, for Rohm and Haas customer William Amer Company, which ran in 1909. That same year, Haas, shown opposite top (second from left), began to record the details of his visits to his Oropon customers and prospective customers in a small notebook, opposite bottom.*

A German Brand in America

Rohm and Haas had good reason to put down roots in Philadelphia. German kid makers liked Oropon, and Haas hoped their American counterparts would too. Trade journals reported the robust state of the fine leather factories in the Delaware Valley, indicating that the time was ripe for Oropon—and for a German chemical company that knew about tanneries.[20]

Otto Haas's new office on 202 North Second Street sat on a trolley line, amid stores, workshops, commercial buildings, and surviving small row houses. In upstairs lofts accountants bent over giant ledger books, while the many small manufactories produced an endless variety of goods for retail, wholesale, and export markets. This oldest part of the city formed a bustling commercial district. A few blocks to the east longshoremen loaded and unloaded cargo at wharves on the Delaware River, where steamships docked from all over the world. Chestnut Street, where banks, importers, auctioneers, wholesalers, and brokerage houses traded, was a short walk to the south.

Philadelphia's steady growth, great size, and diversified economy made it America's preeminent manufacturing hub. Small, enterprising firms could do well, creating innovative, useful products. In textiles alone Philadelphia had seven hundred companies

employing nearly sixty thousand people. On the northeast edge of the city in Bridesburg the chemical plant of Charles Lennig and Company produced commodities like sulfuric acid, used to make the steel cables for the Brooklyn Bridge. In Fishtown, A. J. Reach's sporting-goods factory used local leather in its baseballs, footballs, and boxing gloves. The billowing smokestacks, the cacophony of steam engines, and the odor of the tanneries were the sights, sounds, and smells of progress, making Philadelphia "the workshop of the world."[21]

Otto Haas loved this fast-paced, entrepreneurial environment. In his small rented space he set up a store, an office, and a workroom where imported Oropon was packaged for local customers. He bought himself a large desk in solid oak, big enough for spreading out papers. Over the years he developed a sentimental attachment to the desk, a reminder of the firm's humble American beginnings. (Haas took it to his new offices as the growing enterprise moved to more capacious space on nearby North Front Street in 1912 and then to much grander accommodation still on Washington Square in 1927.) Once settled, Haas hired Germans like Theodore Wenzel, a local tannery worker, who mixed, packaged, and shipped the Oropon orders. Wenzel stayed by Haas's side for decades, eventually becoming a plant manager.[22]

Although in America, Haas was backed by the power of a recognized German brand. Trademarks and brands were new in his day, and these intangibles helped businesses establish their reputation. A firm could attract buyers and hold their attention with a name that was memorable, recognized, and unique. Brand names and trademarks helped make first-time purchasers into repeat customers, establishing trust in the maker and his goods. While Haas's preferred method was to get to know people face-to-face, the brand cemented the relationship. Other German chemical companies had paved the way. American manufacturers knew that the Germans excelled in chemicals, and they wanted German names on their purchases. A trademarked product like Oropon, labeled with the Deutsch name Rohm and Haas, demonstrated quality and reliability.[23]

Calling on the Customer

Haas enjoyed making sales calls, and in the early days he personally handled all the prospects. "In 1909 and the early part of 1910," he recalled, "I devoted my entire time to the tanners using dog manure, located mostly in Philadelphia. . . . I had this field all for myself." By late 1910 and early 1911 he "started to pay more attention to the tanners using hen manure."[24] His approach was straightforward and depended on his ironclad determination to succeed. He would go to a tannery, introduce himself, and ask to demonstrate Oropon. Haas cut an imposing figure, at six feet tall and dressed in a dark gray or black office coat and hat. He talked softly, in slow, thoughtful sentences colored by a thick German accent. Back in his own office he would always say "this is America" and insist that everyone speak English.[25] He practiced what he preached: his notes about the day's work were written in English.

If a tannery agreed to try Oropon, Haas returned to oversee the trials. He described his calls to Charles

Beadenkopf's Wilmington tannery, which typified his approach. "We conducted these experiments with Oropon in order to find out which quantities . . . in what time, and at what temperature would obtain the best possible results." Customers could have figured these things out themselves, but Haas had a strong service ethic. By sharing "my advice and experience" with a customer, he explained, "I save him from loss of time and money."[26]

Haas had enormous respect for the workers in the plant, the practical men with years of experience. Leather making was as much a craft as a science, and there was no substitute for learning by doing. As Robert H. Foerderer of the behemoth Foerderer tannery had written in 1900, "the production of leather is not producing a piece of cloth or iron or anything of that sort. Every skin requires a different treatment. . . . Skin is . . . a natural product, and a difference exists in skins as in men." Haas also encountered and sought to dispel lingering superstitions. In one tannery he sat up all night on the edge of a bate paddle as lightning crackled outside. The beam-house crew taunted him with the folk wisdom that a bating solution would go bad in a thunderstorm. He sat patiently, waiting to prove that Oropon was based on science, not folklore.[27]

Making the sale took great persistence. Tanneries were accustomed to speed in their quality-control tests, but there was no urgency when it came to trying out new materials. In 1909 Haas persuaded a dubious Kaufherr and Company in Newark, New Jersey—a firm that turned horsehide and calfskin into leather for shoe uppers—to try Oropon. In 1913 the tannery was still refusing to give up hen manure on the grounds that Oropon was too expensive. A year later a new superintendent started another run of experiments, which were still going on in 1916, when the notebook entry for Kaufherr stops.[28]

Otto Haas's persistence is displayed in his dealings with another customer, Berkowitz, Goldsmith and Spiegel, which had a goat- and sheep-leather plant in Newark. Sales fluctuated as the tannery contended with unpredictable markets. In 1917 Berkowitz tried different grades of Oropon on cheap, good, and high-grade sheepskins. The next year the firm shifted to glazed kid, running 150 to 200 dozen goatskins a day, using Oropon AB. A year later they dropped goat, cow, and horsehide and went into skivers, a soft, thin leather produced from the grain side of a sheepskin, sometimes using Oropon E and F. When the postwar recession hit in 1920, Edgar Kirsopp, a young Scottish attorney whom Haas had hired and who was to become his right-hand man, visited the Berkowitz tannery and described their bating rooms as "practically shut down." Berkowitz ran a few horsehides in late fall but shut the beam house in 1921. That November, Kirsopp called again and found "business getting somewhat better." As spirits rose in 1922, Berkowitz experimented on "beaver calf as made in Germany," and asked Rohm and Haas for advice. The roller-coaster ride continued for the entire decade. At prosperous moments Oropon orders poured in. When sales lagged, Berkowitz turned to skivers, which did not always need bating. If the price of hides soared, the tannery closed the beam house.[29]

Sometimes Haas took advantage of rivalries among tanners, playing leather men off each other. In December 1913 he rode the train to Wilmington, calling on F. Blumenthal and Company, which had been a customer for a few years. Superintendent Fred Blatz happily reported that he was daily producing a pack of Cabrettas (a light soft leather made from the skins of hairy sheep), achieving fine quality with Oropon. Haas pitted Blatz against the Newcastle Leather Company, a customer that produced several packs per day. When Blatz ventured that Charles Beadenkopf Company had the "best Chinese goat in Wilmington," Haas quietly revealed that they were all bated with Oropon. The news impressed Blatz.[30]

By 1913 the fledgling American firm of Rohm and Haas had no fewer than 120 customers and had especially close relationships with local users. Loyal customers like the Wilmington tanners helped Haas learn about his competitors. That year Fred Blatz gave Haas samples of a synthetic bate sold by Robert Harkinson and Company of Philadelphia. Suspecting an infringement of his U.S. patents, Haas shipped these samples to Darmstadt, where Röhm analyzed them in the lab. Soon Darmstadt cabled the feared news that Harkinson's product was one of several new bates to show traces of pancreatic enzymes. Haas contacted his attorney. A later, bigger worry was Puerine, made by the Martin Dennis Company of Newark, New Jersey, which Rohm and Haas sued in 1916. Here Charles Beadenkopf, who bated twelve thousand skins per week with Oropon, helped Haas, ordering Puerine from Dennis and shipping the barrel to Philadelphia.[31]

The Trials of War

A far greater worry emerged as the European war, begun in August 1914, increasingly threatened to draw in the United States on the Allied side. The British naval blockade of the North Atlantic had quickly cut off supplies of German chemicals, sending American industry into a tailspin. Textile mills and converters suffered for want of dyes and intermediates, and many pharmaceuticals disappeared from hospitals, doctor's offices, and drugstores. Firms like E. I. du Pont de Nemours and Company, mainly an explosives manufacturer, took a crash course in organic chemistry and scrambled to make substitutes for German imports.

Otto Haas separately incorporated the Rohm and Haas partnership in the United States as an American business, just as the United States entered World War I on the Allied side in April 1917. Oropon itself was already a part of the Allied (and also the German) military effort. Early in 1914 Haas had rented a factory in Chicago, the meatpackers' hub, to cope with burgeoning demand for Oropon by using local cattle pancreases. An American chemist with a Ph.D. from Munich, Charles Hollander, got the plant running with help from Karl Stutz, a German chemist from Darmstadt, shortly before the assassination of Archduke Franz Ferdinand of Austria sparked the chain reaction that led to war. Initially both the Allied and the Central powers quite naturally turned to the neutral United States for tanned skins and leather products.[32] Chicago-made Oropon found applications in heavy leather for knapsacks, combat boots, harnesses, saddles, and other military gear. In a way that would become all too familiar as the twentieth century progressed, war could be good for new science-based businesses. Haas quickly purchased farmland in Bristol, near Philadelphia, where he erected a new Oropon factory and an up-to-date research laboratory. The Bristol plant allowed him to close the makeshift Chicago quarters.

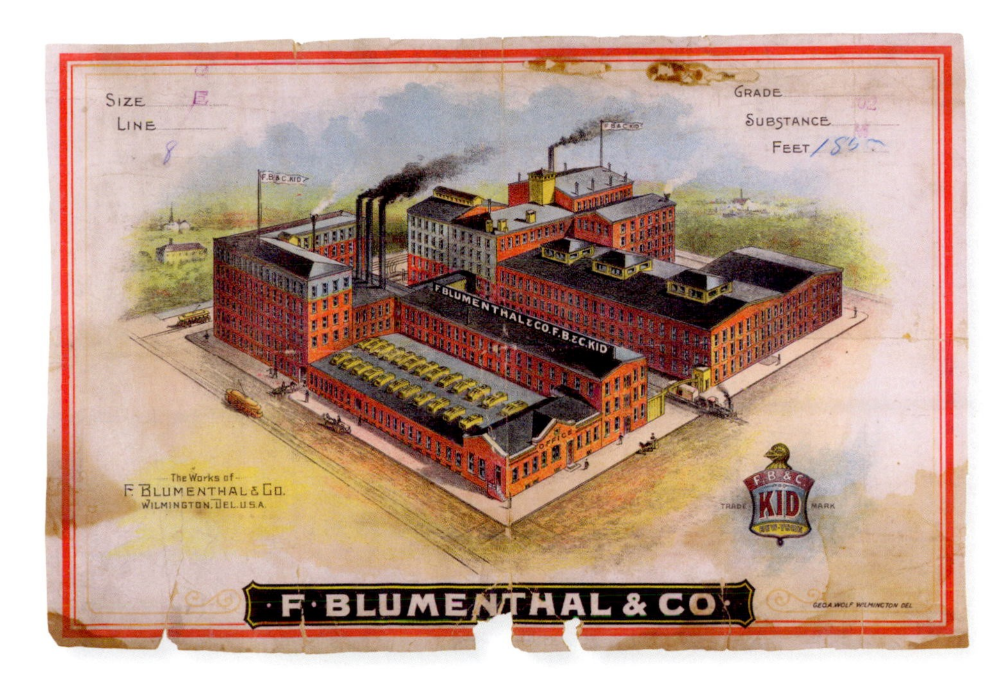

The extensive kid-leather works of F. Blumenthal & Company, above, an early Rohm and Haas customer, covered an entire city block in Wilmington, Delaware. Luxury department stores promoted American kid-leather gloves to customers who wanted European style at a reasonable price, as in this advertising postcard from Marshall Field & Company in Chicago, opposite bottom. Oropon, shipped to tanners in large barrels, opposite top, helped the booming American kid-leather industry by modernizing the bating process.

ROHM AND HAAS COMPANY | **TODAY**

PREPARED AND PROTECTED

Safety is a top priority at Rohm and Haas. At the company's Bristol facility, a full emergency response unit is maintained on site. Here, emergency services technician and shift fire chief Chris Harvey calibrates equipment as part of his routine maintenance procedure.

Building Bristol

In the summer of 1916 Otto Haas and his wife, Phoebe, were acting in ways typical of a young married couple. They had first met in 1913 on board ship, as Otto headed to Argentina and Chile to explore markets for Oropon. Phoebe Waterman in her turn was an astronomer, a newly minted Ph.D. from the University of California, Berkeley, heading south to observe the stars. They were married on February 22, 1914, and were now together for Sunday-afternoon excursions. Distrusting the newfangled horseless carriages and wishing to explore widely, they made full use of the excellent network of steam train and electric trolley lines that radiated out into the countryside around Philadelphia.

Characteristically, Otto Haas was seeking to advance his business, even on a Sunday-afternoon trip. His aim was to visit nearby farm communities and find the best site for the purpose-built Rohm and Haas chemical plant that he envisaged. On an emergency basis he had opened an Oropon plant in Chicago, close to the slaughterhouses, to cope with burgeoning demand as a European war loomed. Now he had decided to make his bate closer to home, where he could directly supervise operations.

Like other Progressive-era businessmen, Haas recognized the advantages of a "green field" site outside the city, with plenty of space for worker housing, research labs, production facilities, and even a farm. The Sunday jaunts paid off when he and Phoebe found the ideal tract of land, in Bristol, a town twenty miles north of Philadelphia. The site had access to the Pennsylvania Railroad and the Delaware River, a major shipping channel. Construction soon began, and Bristol produced its first batch of Oropon in December 1917.

The Bristol plant enabled expanding production and the building of a loyal workforce. Harry Eckert, who in 1924 at age seventeen started as a Bristol lab boy, retired as a research scientist in 1973. "There were only 158 employees in the company when I started. The company educated me, paying for my train tickets to night school at Temple University in Philadelphia. There was a 'trash fund'—money from selling scrap—that was set aside to pay for my education."

In 1927 the Bristol social club was started by Charles Hollander, a vice president in charge of the research labs. Eckert recalled those early days: "Dr. Hollander used to play bocce [a type of Italian lawn bowling] with the guys at dinnertime in front of Building 5. Some of them couldn't speak English. The company hauled in sand to make a beach that extended from the Clubhouse to the Bristol Bridge. I went swimming there many times. There was a small golf course. They used to take pictures at the picnics, using one of those big cameras that you wound up. Later on, they would bring the whole company to Bristol for clambakes, in trucks from Lennig and in buses from the city office. There was never any class distinction between the bosses and the workmen. It was like a big family."

A freight train carrying barrels of Oropon passes by the office-laboratory at Rohm and Haas's Bristol plant, which produced its first batch of Oropon at the end of 1917. Otto Haas chose the site of the new plant, located twenty miles north of Philadelphia, for its access to a major railroad and waterway.

Although business was growing, life on the home front was becoming difficult for Haas and other German immigrants, especially when the United States joined the conflict. Popular culture depicted the German enemy as barbarians and the kaiser's soldiers as heartless Huns who had raped Belgian nuns and pillaged northern France. Newspapers printed sensationalist stories about poisoned bread from German bakeries and stealth U-boats that targeted U.S. ships. In a show of 100 percent Americanism food vendors renamed frankfurters hot dogs, while sauerkraut became liberty cabbage.

Haas's German heritage, once an asset, had turned into a liability. Soon the U.S. Congress passed the Trading with the Enemy Act and created the Office of Alien Property to manage the U.S. assets of German citizens. Following the Armistice in November 1918 lawmakers strengthened the powers of the Alien Property Custodian, granting it authority to confiscate German assets and sell them to Americans. German patents, including those for Oropon, went on the auction block. The aim was to discourage German chemical companies from rebuilding their presence in the United States and to give competitive advantage to American firms that had ventured into chemical research during the war.[33]

Persistent probing from the Alien Property Custodian saddened Haas, now an American citizen who had to defend his loyalty to his adopted homeland. The Tanners Council, an umbrella group for leather makers, came to his aid. Ultimately the organization created the Tanners Products Company, which bought Röhm's assets from the Alien Property Custodian. By early 1920 Tanners Products and Haas co-owned the U.S. firm, and the two Ottos could reestablish their harmonious relationship.[34]

Beyond Oropon

As World War I transformed American industry, it created new challenges and opportunities for Rohm and Haas. Oropon's success was a double-edged sword. Haas began to worry about his dependence on a single product. The firm lost an important patent case in 1918, opening the bating field to competitors.[35] Stanton Kelton, Sr., another young attorney who worked for Otto Haas, later remembered how "from the time I first went into the business in 1916, the one point Mr. Haas kept emphasizing all the time was that a business which is a one-product business is in very great danger of being driven out of business."[36]

In 1918 Rohm and Haas manufactured Oropon in a rented factory, above, that was close to Chicago's slaughterhouses, which provided a steady supply of cattle pancreases. By 1919, when the silver cup below was presented, the company had opened its plant in Bristol.

In 1920 Haas launched a new strategy—one that would come to characterize Rohm and Haas over the years. He sought to diversify the company's offerings by moving far outside the leather industry. He boldly bought Charles Lennig and Company, an old Philadelphia manufacturer of basic chemicals with a dockside plant in Bridesburg, a gritty industrial district seven miles upstream on the Delaware River. Displaying his entrepreneurial energy and eye for opportunity, he had already launched a sideline years before, purchasing sodium sulfide from Lennig for sale to Latin American tanners. With the addition of Lennig as a wholly owned subsidiary, Haas positioned his firm to look beyond Oropon, as the market for synthetic bates matured.[37]

Ultimately Haas agreed to most of BASF's demands, thereby taking steps to leverage his firm's reputation in leather chemicals and to reduce its reliance on the maturing Oropon brand. In the postwar order, with science and innovation moving to the forefront of American industry, he needed products that could expand his firm's share of the leather business.

Otto Haas's first move outside the leather industry came in 1920 when he purchased Charles Lennig and Company, opposite, a basic chemicals manufacturer. Below: Lennig had hired a number of female workers during World War I.

His worries deepened when European tanneries recovered from the war and began to challenge American dominance of kid leather. He watched the nimblest local tanners cope with the 1920–21 recession by shifting to fresh colors, while others foundered, wrestling with shortages of skins, endless style changes, and their declining position in the world market. In an ironic twist of fate Germany was finally combining its fine leather-working skills and its advanced chemical expertise to become a major player in the global kid-leather trade.[38]

The stage was set for serious conversation when Carl Immerheiser, a distinguished German chemist, called at the North Front Street office in April 1921. The previous summer, on one of what became many European trips, Haas had stopped at the leading German firm of Badische Anilin- und Soda-Fabrik (BASF) in Ludwigshafen to discuss changes in the American leather industry. The German chemical giant responded by sending Immerheiser, a top researcher, to talk about Ordoval and Neradol, new BASF synthetic tannins. BASF distributed these "syntans" in Europe but not in the United States, where the Alien Property Custodian had seized its subsidiaries and patents. As an American corporation with strong German roots, Rohm and Haas could be the ideal licensee, especially given its position in leather chemicals. Haas acknowledged Ordoval's promise but winced at BASF's terms: $50,000 cash and a 25 percent commission on sales. "I don't think that they expect themselves to get it," he told Van A. Wallin of Tanners Products. "It is just a question of coming down to brass tacks."[39]

Ultimately Haas agreed to most of BASF's demands, thereby taking steps to leverage his firm's reputation in leather chemicals and to reduce its reliance on the maturing Oropon brand.[40] In the postwar order, with science and innovation moving to the forefront of American industry, he needed products that could expand his firm's share of the leather business. By now Haas had a small cadre of his own chemists who were struggling to develop the possibilities, even as they provided technical service to customers. Their hoped-for syntans were poised for trials at the Elk Tanning Company in Ridgway, a western Pennsylvania plant owned by the giant Central Leather Company. However, Haas suspected that his own product would not measure up to BASF's. "There can be no doubt that they have devoted more time and money, and also have employed better talent on the study of this question than anybody else," he confided to Wallin. "Their products have great merit."[41]

Otto Haas, the master of customer service, was now on the receiving end of technical sales. In late May, Kirsopp sailed for Europe and the BASF factory in Ludwigshafen. He stayed for nearly a month, studying the "manufacture and the experimental work in the laboratory" and touring German tanneries "where the Ordoval . . . is used on a large scale." Haas meanwhile

2
3
4
5
T.C.W.X. 3
C. LENNIG
& CO.
INC.
BRIDESBURG.
PA.
T.C.W.X.

In 1921 Otto Haas began experiments on the Ordoval syntan at Central Leather's tannery in Ridgway, Pennsylvania, above, and the Max Hertz Company, which made automobile leather for cars like the Hudson-Essex featured in the ad at right.

cautioned that "until we have our manufacture straight and until we have actual experience in a tannery, I think we ought to keep rather quiet on Ordoval."[42]

This approach was only prudent, as Wallin, on a tour of Newark tanneries, learned from his friends in the leather trade that the Ridgway people were "quite fearful of any extensive experimental work with synthetic tannins. . . . They feel that they know a good deal about them, and their experience does not make them hopeful of any large use . . . in heavy leather." "They all believe that only after a long series of small experiments would they undertake any large operations. . . . Their conservative general policy" will make them "a rather unreceptive field for rapid experimental or demonstration work." Two days later on the train from Newark to Boston, Wallin sat with two other leather men who seemed more interested. From his room at Boston's Hotel Bellevue, he wrote Haas: "You may get better results at other places than Ridgway."[43]

By the fall, experiments on BASF's Ordoval syntan, renamed Sorbanol, were under way in several places, including Central Leather's Tionesta Tannery in Sheffield, Pennsylvania, and the Max Hertz Company in Newark, New Jersey, which made automobile leather. Haas was still cautious, preferring "to get the necessary experience in a limited number of tanneries before placing the product for sale at large." The most important experiments were at Central Leather, a company created in 1905 from a consolidation of the United States Leather Company and several smaller firms. Central Leather was an altogether larger customer, and Haas was acutely aware that large orders might roll in from the "Leather Trust."[44] Inexpensive American shoes dominated the world market, and the big sole-leather tanneries devoured a huge volume of material, making them a terrific potential customer for Rohm and Haas.

Selling to big businesses was new terrain, and Otto Haas had to feel his way. His German name, which had been an asset in the prewar kid-leather trade, did not carry much weight with American industrial giants. The Chicago tanneries affiliated with the big meatpackers were shrewd and cautious. Haas approached M. P. Brennan, general manager at the Armour Leather Company in upstate New York, only to be rebuffed. At headquarters in the Windy City president H. W. Boyd had heard that leather made with the new synthetic tannins deteriorated over time. The rumors made Brennan wary.[45] When Armour finally tested Sorbanol, Haas believed that "the material was only tried out in a half-hearted way."[46] The trade gossip on syntans reached George W. Johnson, head of the Endicott Johnson shoe empire, who asked probing questions before finally giving Sorbanol a trial run.[47]

Slowly Rohm and Haas introduced Sorbanol to the large sole-leather tanners. Some even experimented with newer BASF syntans, including Leukanol, a variation on Sorbanol.[48] In 1922 alone Rohm and Haas salesmen and chemist-demonstrators tested Sorbanol in seventy-five American tanneries, obtaining a "great deal of valuable, practical experience as to the possibility and limitations of this product." Customers included "tanners of all kinds of light and heavy leather, sheep, calf, horse, seal, pack and case, insoles, automobile, chrome retanned, and sole leather."[49] The company's business was moving far beyond its origins with Robert H. Foerderer and other kid leather customers of the early years. Otto Haas was both making money and raising his sights.

The strategies of Rohm and Haas were characteristic of how the small American specialty-chemical industry worked in the peace that followed World War I. A complex web of interpersonal, intercompany, and international networks linked dye makers and textile mills, chemists and tanners, Germans and Americans, big business and small. The Alien Property Custodian may have wanted to expel all German firms from the United States, but giants like Hoechst, Bayer, and BASF had too much chemical expertise to disappear from the commercial scene. American companies could no more exist without German technology than the vici tanneries could survive without Asian goatskins. This interdependence made zealous patriots uncomfortable, but thoughtful executives like Haas had better insight into the commercial realities.

An Emerging Enterprise

Over the course of two decades Otto Haas learned much about innovation through collaboration and salesmanship through technical expertise. His focus on the customer complemented the talents of his friend Otto Röhm, the science genius, who tinkered with molecules to solve the mystery of bating. After first introducing Oropon to German tanneries Haas had returned to the United States, his chosen place to live, to win over Philadelphia. As he showed the local factories why Oropon was better than dog dung, he saw how American tanneries operated. His insights into leather chemistry were freely shared with practical men, and in turn they passed on their knowledge of skins and hides, prices, and Rohm and Haas's competitors. Soon, patented innovations and hard-won trade secrets were translating into steady sales and growing profits. Ambitious and shrewd, Haas plowed profits back into business and growth.

While Otto Haas continued to look for sales opportunities from his base in America, Otto Röhm held down the partners' German operations. Röhm, wearing an apron over his business suit, oversees workers as they dehair skins in the experimental tannery at Darmstadt, above.

After the war disrupted this equilibrium, he searched for the ballast to steady his company in the new environment. Philadelphia kid lost its grip on world markets, but enough U.S. tanneries used Oropon to keep Rohm and Haas in the black and to maintain the profits that enabled innovation. In 1926, 20 percent of sales and 53 percent of net profits still came from Oropon. Other major leather chemicals, such as Sorbanol and Leukanol, were 5 percent of sales and only 3 percent of profits.[50] Haas's prescient observation about the hazards of focusing on a single business was playing out. The route forward had to be through diversification. One answer lay in the purchase of a company like Lennig and the invigoration of its sales. Another possibility lay in finding other new chemicals through research and applying them to the novel needs of fresh customers. With these thoughts in mind Haas looked abroad. His firm's German beginnings and continued links simultaneously indicated and enabled this strategy in the 1920s and 1930s.

02 | TRANSATLANTIC CONNECTIONS

Positioning for Growth

On a Saturday morning in May 1927, Otto Haas pulled up stakes. The 40 North Front Street offices, adequate in 1912, had become crowded as the business expanded and the staff grew to forty people. When the Philadelphia Gas Company put its office building up for sale, Rohm and Haas seized the opportunity to buy their elegant five-story brick structure at 222 West Washington Square.[1] The new location brought the company into proximity with some of Philadelphia's leading enterprises. Washington Square was home to Curtis Publishing, known for the *Saturday Evening Post* and *Ladies' Home Journal*; J. B. Lippincott, the venerable Quaker publishing company; and a handful of other book and magazine publishers. Nearby stood N. W. Ayer and Sons, a top advertising agency, and the Philadelphia Savings Fund Society, the country's first savings bank. Otto Haas's new

THE WASHINGTON SQUARE BUILDING, OPPOSITE, SERVED AS ROHM AND HAAS HEADQUARTERS FROM 1927 TO 1965. OTTO HAAS HAD THE SECOND-FLOOR CORNER OFFICE. DURING WORLD WAR II THE COMPANY PROUDLY FLEW TWO FLAGS: OLD GLORY AND AN ARMY-NAVY "E" BANNER FOR EXCELLENCE IN WAR PRODUCTION. PREVIOUS SPREAD: IN THE 1930S ROHM AND HAAS'S CHEMISTS DEVELOPED THE UFORMITE RESINS USED TO ENAMEL APPLIANCES, MAKE PLYWOOD, AND PRODUCE STRONGER PAPER. IN 1951 THE HAWTHORNE PAPER COMPANY USED UFORMITE 700 TO MANUFACTURE MAP PAPER THAT HELD TOGETHER EVEN WHEN DRIPPING WET.

Otto Haas's double desk, above, sits in his sparsely appointed room overlooking the park. He had purchased the desk for his first office and maintained a sentimental attachment to it, taking it with him as the Rohm and Haas offices moved and grew.

office, equipped with a large fireplace, overlooked the leafy square, one of the five parks in William Penn's 1682 visionary plan for Philadelphia and a burial site for Revolutionary War soldiers.[2]

The move to Washington Square coincided with the transformation of American culture after World War I. In 1917 the Victor Talking Machine Company released the first record of the Jazz Age, launching a party that ended only with the stock market crash of 1929. Women had the vote, young men were off to college, and backseat smooching replaced front-porch spooning. Yet the ebullience hid discontent. Prohibition, which outlawed alcohol from 1920 to 1933, gave birth to organized crime, bathtub gin, and speakeasies. Jazz bands and flapper fashions were bright anomalies in a decade that was fundamentally conservative and inward looking.

In 1925 President Calvin Coolidge coined a phrase that described the prevailing mood: "The chief business of the American people is business." Three successive Republican administrations—Harding's, Coolidge's, and Hoover's—gave business a loose rein. In a backlash against trust busting the Justice Department turned a blind eye to monopolies, allowing corporations like Central Leather, U.S. Steel, and AT&T to become industrial giants. Congress passed legislation that permitted collaboration with European cartels and the cozy division of world markets. A new Alien Property Custodian took office, more favorably disposed toward German firms and their American business associates.[3]

This new economic order was tailor-made for the persistence and drive of Otto Haas and his itch to expand the business of his innovative and very profitable chemical company. The era's unbridled consumerism favored alert manufacturers of chemical specialties. Flapper fashions, for example, gave a boost to New York's garment makers and their textile suppliers, who used chemicals in processing colorful new fabrics like rayon. Haas boldly ventured into this new territory of textile chemicals, reaching for German technology to move him beyond Oropon as his only major product. In due course Rohm and Haas made substantial profits from compounds like BASF's Lykopon and Formopon, hydrosulfites used in coloring fabrics with the new vat dyes. By 1929 textile chemicals provided no less than 40 percent of gross sales, while Oropon accounted for a mere 14 percent. Other promising lines that Haas took up included organic insecticides and cuprous oxide, an ingredient in antifouling marine paint for U.S. Navy ship bottoms. The golden age of leather had passed, and the era of diversification began.[4]

If Haas had a disappointment, it was in overseas operations. Initially and ambitiously, he and Röhm envisaged selling Oropon on five continents: Europe, North America, South America, Asia, and Australia. Resident agents or traveling salesmen would represent the partners in different territories. Besides North America, Otto Haas had Latin America, New Zealand, Australia, and Japan as his territories, while Röhm had the lion's share, from Europe to India. Already in 1913 Haas had visited Latin America to explore the market. Sales in Latin America expanded during World War I but dropped precipitously in the 1920s as the German chemical industry recovered and resumed its aggressive marketing. Haas closed his sales offices in Argentina and Chile, ruefully confiding to Röhm: "I sometimes thought of handing South America back to you."[5]

In his Washington Square office Haas reflected on his accomplishments and pondered the future, as the 1920s drew toward its close. He had steadily diversified since the war, acquiring the U.S. patent rights for various German products. He had also developed human capital, expanding research facilities at the plants and opening laboratories on the top floor at Washington Square.[6] Gradually Rohm and Haas was acquiring the ability to move in many directions beyond Oropon. But Haas, who had watched Röhm at the bench, knew that his own industrial research would not yield results overnight. He needed more and better technology, sooner rather than later. For answers he looked still more determinedly beyond Washington Square, to his German homeland.

Getting the Know-How

There were two ways to get chemical know-how: develop research internally or acquire it externally. After World War I, industry leader E. I. du Pont de Nemours and Company (DuPont) did its best to master synthetic organic chemistry, which had given Germany the edge in dyes, fertilizers, and explosives. In a scenario direct from a spy movie DuPont hired and smuggled out of Germany four leading dye chemists in a bid to jump-start its Wilmington-area operations. Most strategies were less cloak-and-dagger. DuPont also signed an agreement with Great Britain's Imperial Chemical Industries to share scientific and technical information, with the firms granting each other exclusive licenses for patents and processes and working around antitrust laws to divide world markets.[7]

Dow Chemical, in its turn, obtained licenses to manufacture dyestuffs from the Swiss company Ciba, which had cross-licensing agreements with German firms. Other American companies went directly to Germany. Despite Germany's defeat its chemical industry was still a reservoir of talent, its factories and research laboratories the envy of chemists throughout the world. General Electric collaborated with German firms to create resins for the electrical industry, while DuPont signed contracts with I.G. Farbenindustrie, or "the I.G."—the giant trust created by the 1925 merger of BASF, Bayer, Hoechst, and three smaller firms—to share patents and processes.[8]

With eyes fixed on Europe, Haas was no different from other American chemical manufacturers. To take advantage of burgeoning U.S. markets, outsmart the competition, and sustain the growth that began during the war, he had to find new science and new technology. In forging relationships with German chemical companies Haas, thanks to the advantages of birth, was well placed. His knowledge of the language and culture and his strong ties to Otto Röhm opened the doors at major firms. For example, Haas established a friendship with Carl Bosch, BASF's managing director and a future Nobel laureate in chemistry, which triggered an important connection between Philadelphia and Ludwigshafen.[9] Most summers through the 1920s and 1930s Haas and his wife, Phoebe, would spend six to eight weeks in Europe, sailing from New York on a luxury liner like the Cunarder Laconia, splitting their

As Otto Haas sought to diversify his company's product line at the end of the 1920s, he looked to his homeland of Germany. The Resinous Products and Chemical Company, an American firm Haas established with Frankfurt chemist Kurt Albert in 1927, had a varnish laboratory, above, on one of the top floors of Rohm and Haas's Washington Square headquarters.

Initially, Resinous Products continued the characteristic modus operandi established by Otto Haas before the war, adapting German inventions to the American market by working closely with customers. The Bakelite threat pushed Haas to vigorous development of original research in the United States to complement Albert's work in Germany.

Pratt and Lambert's quick-drying varnishes, advertised in the Saturday Evening Post *in 1926, above, were emblematic of the pace of the Jazz Age. Otto Haas recognized the opportunities presented by these types of products and hurried to respond.*

time between the Upper Rhine and the Seine and combining business in Germany with sightseeing and art collecting in Paris.[10]

These working vacations gave Haas a welcome respite from his intense Philadelphia routine of paperwork and plant visits. They also refreshed his friendship with Otto Röhm and scientists in the Darmstadt labs and enabled his connection to the I.G.'s state-of-the-art facilities in Ludwigshafen, Düsseldorf, and Frankfurt. Old ties blended easily with new technologies and fresh business partners. Rohm and Haas, conceived as an international venture, was well positioned for a business environment in which better things for better living through chemistry was both a promise and a daily reality.

Fast Finishes

In the early 1920s DuPont launched a coatings revolution with Duco, a quick-drying lacquer that General Motors put on its "True Blue" Oakland car. A nitrocellulose product, Duco dried in a mere four hours, shortening from days the time needed to paint a car and spurring a wide interest in better paints and varnishes. By mid-decade the coatings industry offered similar products to furniture companies, master painters, homeowners, and housewives. Colorful ads in the *Saturday Evening Post* targeted do-it-yourselfers with quick-drying paints and varnishes, which appealed to the Jazz Age wish for style and speed.[11]

The fervor over fast finishes quickly caught the attention of Otto Haas, who met the chemist-entrepreneur Kurt Albert in Frankfurt in 1924. Albert's firm made Albertol, a line of phenol formaldehyde resins that replaced the tree sap in varnishes. Rohm and Haas tried selling imported Albertol in the United States, but varnish makers showed little interest until nitrocellulose lacquers took off and made them realize the future lay in fast-drying coatings. The Haas-Albert synthetic resin, renamed Amberol, proved unsuitable for nitrocellulose lacquers but instead became the oil-soluble resin of choice for other types of varnishes.[12]

Amberol found a ready market in shelf goods that lumberyards, paint stores, and hardware dealers marketed to homeowners, handymen, and master painters. Haas and his new partner, Albert, combined resources to set up a new American firm, the Resinous Products and Chemical Company, which opened for business in January 1927. Legally and operationally separate from Rohm and Haas, Resinous Products nonetheless shared offices at Washington Square and research and production facilities at Bridesburg. Haas handed the new company's day-to-day operations over to Edgar Kirsopp, the lawyer who had been by his side since 1919. Amberol enjoyed a clear success when George D. Wetherill Company, a leading Philadelphia paint manufacturer, put it in a new interior varnish that dried rapidly, allowing the painter to apply the two requisite coats in a single day. By late 1927 three large national manufacturers—Valentine and Company, Boston Varnish Company, and Pittsburgh Plate Glass Company—adopted Amberol, encouraging still others to try the new resin.[13] Its acceptance led Haas and Kirsopp to expand the new company's portfolio of resins.

"Within recent years, interest in synthetic resins, not only for moulding, but for lacquers and varnishes, has grown enormously, and several large firms and innumerable small concerns are now engaged in research development in this field," Haas wrote in 1927. Competitors included the Beck, Koller Company and the Paramet Chemical Company, each of which introduced imitation Amberol under trade names like Beckacite and Paranol. Big players, such as DuPont, General Electric, American Cyanamid, and Union Carbide, also pushed new kinds of resins for varnishes and lacquers. The real threat came when the hugely successful General Bakelite Corporation, whose Belgian founder, Leo Baekeland, had invented a synthetic plastic back in 1907, ventured into the territory. When customers started using Bakelite's new XR-254 resin to make durable outdoor varnishes, Haas and Kirsopp got nervous.[14]

Initially, Resinous Products continued the characteristic modus operandi established by Otto Haas before the war, adapting German inventions to the American market by working closely with customers. The Bakelite threat pushed Haas to vigorous development of original research in the United States to complement Albert's work in Germany. In 1930 Haas hired Ph.D. chemist W. T. Pearce, who had been a professor in the paint and varnish school at the North Dakota Agricultural College and had worked for Valentine and Company. Soon Haas was reporting that "as in former years, we have spent considerable money on our research laboratories, because we think it is of the utmost importance to open up new fields of opportunity and to diversify our interests." Pearce set to work on coatings. Over the next ten years Haas and Kirsopp expanded the Resinous research staff to include nineteen chemists in the labs of this separate company, at Washington Square and at Bridesburg.[15]

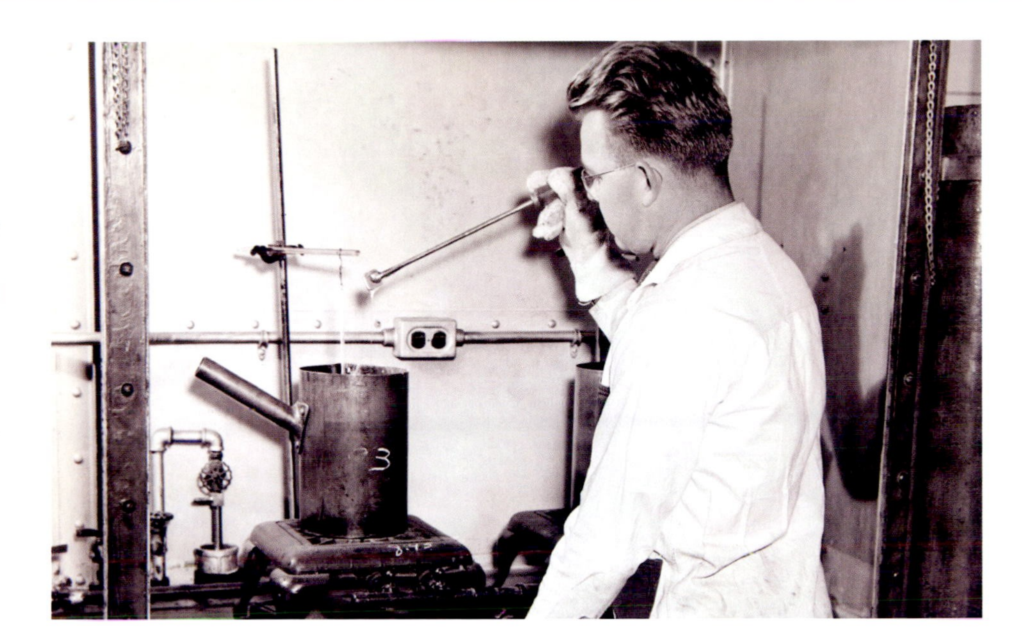

A chemist cooks Amberol, a synthetic resin used in certain types of varnishes, in a Resinous Products lab circa 1945, above. Rohm and Haas's Bridesburg plant, below, housed Resinous Products' research and manufacturing facilities.

Stuck on Plywood

Neither developing products in-house nor acquiring foreign technology and adapting it to the American scene were easy, as shown by the trials and tribulations with a product called Tego Gluefilm. A phenol formaldehyde resin impregnated into wood-pulp paper, Tego was patented by Theodor Goldschmidt Corporation, a German firm with a factory in Essen and a sales office in New York. Introduced in Europe in 1929, Tego Gluefilm was a substitute for liquid glues made from plant and animal by-products. This adhesive was less messy and aged better, making it ideal for furniture veneer and plywood.[16]

ROHM AND HAAS COMPANY | **TODAY**

PLANNING FOR GROWTH

Rohm and Haas maintains more than one hundred facilities in twenty-seven countries. Here, construction services manager Michael Hays (left) and home office building manager Louis Velez review plans for renovations made to the corporate headquarters in 2008.

Resinous Products workers inspect a roll of Tego resin adhesive in the 1940s, above. The resin-adhesives industry was bitterly competitive, which motivated Otto Haas eventually to look for developments outside the field of resins. Right: This advertising booklet explained Tego's advantages to plywood manufacturers.

Kirsopp learned about Tego in 1933, when R. E. Vogel, a scientist for Otto Röhm in Darmstadt, wrote excitedly to Philadelphia. That summer Kirsopp visited the Goldschmidt headquarters in Germany to inquire after Tego's American rights. Managers reported that Goldschmidt already had a U.S. distributor but urged Kirsopp to sit tight because the relationship was on shaky ground. Back in Philadelphia, Kirsopp met with Ray Sorenson, the salesman who introduced Tego to the American market. Sorenson told Kirsopp about ongoing conflicts between Goldschmidt and its American distributor, Tego Gluefilm, Incorporated, that interfered with his ability to sell more Tego. Meanwhile, in Essen, Goldschmidt puzzled over American sales, which paled in comparison to those in Europe. Finally, Goldschmidt canceled Tego Gluefilm's contract and licensed Resinous Products as the North American manufacturer and distributor. Underlining the importance of trade secrets and technical know-how in the world of specialty chemicals, Resinous Products noted that "the only reason we were willing to enter into the agreement was in order to get the benefit of all of the secret information, formulae, manufacturing plans, etc., which Th. Goldschmidt, of Essen, Germany, had developed during the long series of years."[17]

Kirsopp and his staff could now work to establish Tego adhesive in the United States and Canada. While Hans Haeberle, a Bridesburg foreman, went to Essen to study Goldschmidt's production line, the sales staff probed the potential market. Prospective customers, including major plywood, lumber, sporting goods, and furniture manufacturers, smiled favorably on Tego, speculating that it "would revolutionize the wood industry." The New Albany Veneering Company, the country's largest plywood manufacturer, with ties to Weyerhaeuser, thought it might be ideal for prefabricated houses.[18]

Since heat was needed to activate the adhesive, Tego sales were limited to factories willing to install hot presses. Sales were slow but increased steadily for several years. By 1939, however, Haas and Kirsopp had reluctantly concluded that Tego was at best a niche product, likely to find favor only among a small group of plywood and veneer manufacturers. That year 75 percent of Resinous sales, or just over $3 million, was in coatings, but a mere 10 percent, or $400,000, came from Tego plywood adhesive.[19]

Why did such a promising technology stagnate? Newer adhesives made from urea formaldehyde resin, including Resinous's own Uformite 430, gave Tego a run for its money. Uformite liquid adhesive, introduced in 1937, was easier to handle and required a lower bonding temperature, so manufacturers did not need hot presses to use it. The adhesive territory was also getting crowded, with bitter price competition. Despite the Great Depression, Americans continued to remodel homes, buy new ones, and update furnishings. Chemical companies, often described as "recession proof," responded by acquiring German technology or using in-house expertise to develop synthetic adhesives.

Chemist Harold Turley, above, examines leather processed with Rohm and Haas chemicals. Turley headed the company's first leather lab, where he worked on creating a new type of Oropon, and then went on to lead Rohm and Haas's Leather Research Department for thirty-seven years.

Into Leather

Harold Turley loved leather—the many varieties of hides and skins that chemistry could render soft, pliable, and beautiful. During World War I the Londoner had served with the Royal Army Medical Corps in Iraq, where his work in a tropical-disease laboratory piqued his chemical curiosity. The returning veteran secured a scholarship to Battersea Polytechnic, headed by leather specialist Sir Robert Pickard. In October 1924 Turley completed his Ph.D. and on Pickard's urging hopped on a transatlantic steamer to see about a job with Otto Haas.

Haas already employed two Ph.D. chemists—Charles Hollander at Bristol and Siegfried Kohn at Bridesburg—but he needed a full-time organic chemist to improve his position in leather. "I came over here not knowing a soul!" recalled Turley. "I went to 40 North Front Street. It was a ramshackle place, broken down, and I thought, 'What have I done? Is this a chemical company?' Then, this old man (to me, he was old—he was approaching fifty) came out of the office, threading his way through the barrels. It was Mr. Haas."

The two men hit it off. Haas sent Turley to Columbia University's College of Physicians and Surgeons to study how the pancreas dissolved proteins and sugars. It was smooth sailing until 1926, when a French proposition to make bates from a fungus threatened Oropon's market dominance. Haas brought Turley back to Bristol, where he set up the company's first leather lab and synthesized a new type of fungus-based Oropon. "It was very secret—we weren't allowed to talk to each other, the other chemist and me. They wanted to be careful that no information got out."

Eventually Haas consolidated the Bridesburg and Bristol leather labs, creating the Leather Research Department that Turley headed for thirty-seven years. "We had a little experimental tannery at Bristol. When the Bridesburg people came, we had a larger area and put in equipment like the tanneries." The chemists made leather and deliberately ruined it. "If the product deteriorated in the field, we could have lost our shirt."

Despite the steady growth in prosperity, scale, and complexity of his company, Otto Haas continued to favor a penny-pinching style and to exercise vigilant control. Employees were expected to watch their expenses and find ways to improvise. In retirement Turley recalled how "it was a very *sound* company. I never lost a day's pay. Haas was meticulous about those things. Sometimes he made it hard for you because he wanted to reduce costs. I wanted a little hand centrifuge that cost $15 and couldn't afford it. One time, Hollander said, 'Why don't we use jelly glasses instead of chemical beakers? We can get them at Woolworth's.'"

Otto Haas made frequent trips to Germany to visit the plant in Darmstadt and check in with German partner I.G. Farben. The engraving of the Röhm and Haas Chemische Fabrik in Darmstadt below was created in 1920. Opposite: After a busy day at the Darmstadt plant Otto Haas returned to his accommodations in Frankfurt and wrote his trip report on hotel stationery. The pencil mark through the text was made by his secretary after his return to the home office to show that his remarks had been recorded.

American Cyanamid Company, maker of Beetle and Melmac plastics, captured the largest share of the business. The Plaskon Company and the Casein Company both had American rights to I.G. Farben's Kaurit resin. The formidable Bakelite Corporation introduced a phenol formaldehyde resin similar to Tego Gluefilm and a urea formaldehyde adhesive that competed directly with Uformite.[20]

Regionalism and costs were also important. The plywood industry was divided into two branches: hardwood in the East and Douglas fir in the West. Resinous captured the eastern business, including furniture companies comfortable with hot presses, but could not make inroads on the Pacific Coast. Soft and brittle, Douglas fir was made into plywood by large factories in the West concerned with volume and speed rather than quality. For them Tego was too labor intensive. One western firm agreed to try out Tego, installing presses to use the film but then quickly abandoning the time-consuming process. Another company that ventured into resin-bonded plywood simply decided to make its own adhesive.[21]

Haas and Kirsopp learned important lessons from these disappointments. One was that a novel product of itself does not ensure success. Certainly Tego pushed the building industry in a new direction, introducing plywood manufacturers to an adhesive that was waterproof, weatherproof, and fungus-proof. Later, Uformite extended resin bonding to lower-priced grades of plywood. However, both of these products bumped up against a variety of glues that seemed to work just fine or at least well enough to block any changes.[22]

Even Otto Haas's tried-and-true strategy of close attention to the customer could not guarantee results. "We have to recognize that within the field of resins the possibilities of expansion are becoming more and more restricted," he reflected in 1937. "Not only are there an increasing number of large and important companies working in the field of synthetic resins for coating purposes, plastics, etc., but an increasing proportion of the outlet for resins may be taken by types other than those we manufacture. It may be necessary for us to consider to an increasing extent taking up developments outside of the field of resins themselves."[23]

Mehr Deutsche Technik

Itself the largest European corporation through the 1930s, the I.G. was also the world leader in synthetic organic chemistry. In 1930–31 its new headquarters opened in Frankfurt am Main. A massive concrete-and-steel structure, it testified to the combine's power and ambition. South of Frankfurt in the Upper Rhine industrial district stood the I.G.'s prize manufacturing sites, the giant BASF plants at Ludwigshafen and Oppau. These sprawling complexes housed unrivaled know-how in dyes and dyestuffs intermediates, along with the famous Haber-Bosch process for synthesizing ammonia, explosives, and fertilizers.[24] Here the world's best chemists created new synthetics: plastics, paints, tannins, colorants, oil from coal, and man-made rubber. Chemists from other countries looked on in awe.

On his visits to the Upper Rhine, Haas stayed at the Hotel Frankfurter Hof. Its structure epitomized Old World elegance, beckoning visitors with a columned facade that harked back to past glories. After a long day of meetings Haas could relax in a plush downstairs parlor or in the privacy of his room, using hotel stationery to scratch out notes that would be typed into trip reports back in Philadelphia. These reports reveal the intricacies of the relationship between the American and German chemical industries and especially between Rohm and Haas and the I.G.[25]

HOTEL FRANKFURTER HOF

FERNSPRECHER: STADT: 20012, AUSWÄRTS: 20276
TELEGRAMM-ADRESSE: FRANKHOF FRANKFURTMAIN
POSTSCHECK-KONTO: FRANKFURT A.M. NR. 15512

FRANKFURT AM MAIN

ABSENDER IST NICHT DER FRANKFURTER HOF

DEN 20th August 1936

R & H, Darmstadt

Lethane. I discussed the situation with Dr. Röhm. It seems they have done nothing so far mainly due to the fact that the Plexiglas development consumes all their energy and resources.

Dr. Röhm told me that he was afraid that I.G. is taking up Thiocyanate for insecticide use and he referred in this connection to a conversation with Mr. Klingoff. I told him that Mr. K. probably had told him to study his sources of raw material but Dr. R. still thought I.G. may be in the field and asked me to find out, which so far I have not been able to do. Possibly I can do it on my return to I.G. in September.

Darmstadt ~~will~~ may be able to develop a household insecticide; but they will never make an agricultural insecticide.

Burnus. This matter was gone over again and I repeated what we had written on. Their statement that A. positions can take the place of 98 is slightly premature. There are still a number of difficulties. I asked for a written report, particularly as to their requirements for 1937. I told Dr. R. we need this information soon.

Dr.'s propaganda, particularly the advertising in the papers, on posters in R.R. stations etc. is very good. It is the work of one man.

Lecithin. Dr. is not informed on the situation.

While Haas used German technology to stay on top in leather, this market appeared to have little growth potential. . . . The Depression crimped demand for leather. People resoled rather than replaced their shoes, factories did not install new belting, and consumers bought fewer armchairs or suitcases. Even Detroit used less leather upholstery.

With the help of his wife Elisabeth, Otto Röhm developed Burnus, below, a laundry soaking agent based on enzymes. Introduced in Germany in 1914, Burnus was so far ahead of its time that homemakers hesitated to use it. Ultimately, enzymes found widespread use in detergents during the 1960s.

Heavily committed to exports, the I.G. coveted American customers who could import products, purchase formulas, and create a steady revenue stream from sales and royalties. As the memory of World War I faded in the United States, prejudice against the Germans also waned. German firms discreetly reestablished their American presence through sales agencies, private corporations, and licensees. The I.G.'s reservoir of knowledge gave it the upper hand, and small entrepreneurs conceded to the German giant. Haas continued to rely on BASF, the biggest company in the I.G., for technology in three major areas: leather, textiles, and acrylics. Haas had begun to do business with BASF in 1921, when he acquired the American rights to Carl Immerheiser's patented syntans. Haas and Kirsopp often sought Immerheiser's advice on leather topics, also obtaining the license for Orthochrom, his patented nitrocellulose leather finish. Immerheiser himself became a corporate director and head of the tanning department at BASF, within the newly consolidated I.G.[26]

Haas needed the I.G. if he were to stay competitive in leather, develop a presence in textiles, and move into new areas like acrylic plastics. The relationships became steadily more complex, involving the two Ottos, the Philadelphia and Darmstadt firms, cross-licensees like DuPont, and various I.G. offices, labs, and subsidiaries. There were bilateral and trilateral contracts to share patents and channel royalties to Germany, along with an implicit understanding that less-formal exchanges could benefit both sides. An example was the 1929 effort by Rohm and Haas and the General Dyestuffs Corporation, the I.G.'s distributor in New York, to develop leather colorants in pale fashion hues. The two firms informally divided the North American market between them and together promoted BASF syntans. Since his own labs were new, Haas needed these relationships and willingly paid the royalties.[27] On syntans specifically, he noted, "Our laboratory has not been able to produce anything of outstanding merit; on the other hand, the I.G. has developed and given us new materials which represent a very decided step forward."[28] A few years later Haas observed that "we have used our very best efforts to develop the I.G. products in the right direction, and I think we have been fairly successful."[29]

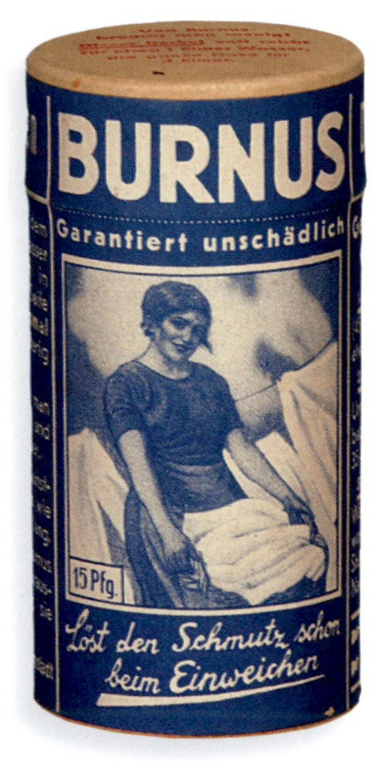

While Haas used German technology to stay on top in leather, this market appeared to have little growth potential. American tanners, with abundant supplies of vegetable tannins, did not rush to the syntans, while mass-market shoe factories scoffed at the idea of a waterproof coating for calfskin. The Depression crimped demand for leather. People resoled rather than replaced their shoes, factories did not install new belting, and consumers bought fewer armchairs or suitcases. Even Detroit used less leather upholstery.[30]

Determined to keep driving ahead in its now familiar role as a customer-oriented specialty company, Rohm and Haas looked to the I.G. for other high-tech products that matched its development, production, and marketing capabilities. A promising innovation from the Darmstadt labs, acrylic plastics, drew Rohm and Haas interest. The I.G. had patented prior art on acrylic resins, leading to cross-licensing by the I.G., DuPont, and Rohm and Haas in Darmstadt and Philadelphia.[31] The 1934 acrylic contract, along with other agreements in alkyd emulsions, leather, and textiles, necessarily entangled Philadelphia and Darmstadt with the I.G.'s own growing accommodation to Germany's new National Socialist government.

As the Depression deepened, the I.G. tightened its belt and carefully managed its royalties, making work for its business associates more cumbersome. A native-born German, Haas always enjoyed wheeling and dealing with German companies. Even so, the I.G.'s bureaucracy was tough to navigate. "Agreements with I.G. are difficult and slow in making," Haas wrote in a trip report. "The technical and legal departments at Ludwigshafen may have drawn up an agreement which is mutually satisfactory; but by the time it reaches the Frankfurt office, it is all off, because one of the officials there may have started something in U.S. or may plan to start something in U.S. of which nobody at Ludwigshafen knows."[32]

The situation worsened at mid-decade when the National Socialist government launched its Four Year Plan for rearmament and self-sufficiency. Building their war machine, Nazi officials pressed the I.G. to speed up projects on synthetic materials. Hydrogenation—the creation of gasoline from coal—and Buna man-made rubber were the most visible of these efforts, which also extended into fibers and plastics. During this transition Haas struggled to fathom changes in the I.G.'s corporate culture and the new tenor of his relationships. On a typical trip in September–October 1936, he visited Darmstadt, the I.G. in Ludwigshafen and Frankfurt, and the representative of a Dutch firm in Paris. At Darmstadt discussions revolved around Burnus enzyme detergent, Lethane insecticides, and developments in acrylics—and Otto Röhm's frustrations with the BASF's tanning products. "They still have a lot of friction with the I.G.," Haas somberly noted, "and I tried to help straighten them out."[33]

A new Darmstadt office building, erected by Otto Röhm, was appointed with etched art-deco windows, left, made from the shatterproof acrylic plastic Plexiglas. Art enthusiast Röhm created the Darmstädter Neue Glaskunst, where artists explored Plexiglas as an expressive medium.

At Ludwigshafen, Haas visited the expanded tanning department—syntans were now his best-selling leather product—and learned about cutting-edge plastics research. Discussions centered on syntans, styrene, plastics, acrylics, and gossip about competitors on both sides of the Atlantic. Haas's requests for more information on styrene were evaded. "I had a good opportunity to ask Dr. Kessler on the possibilities of getting the Styrol rights for the U.S.," Haas scribbled in his report. "He was somewhat vague and merely stated that he thought the patent situation was rather confused. The fact seems to be that such matters are decided by Fritz ter Meer [the top manager] and everybody is careful not to butt in."[34] Haas later learned

ROHM AND HAAS COMPANY | **TODAY**

ADDED IMPROVEMENT

Rohm and Haas develops specialty additives that improve the appeal and effectiveness of household products. Here, research scientist Glenn Derricotte prepares an experiment on a dispersant used in detergents to break down, disperse, and prevent redepositing of soils.

The high-tech plastics laboratory at Darmstadt, shown above circa 1937, was the birthplace of acrylic plastic, a clear, moldable, shatterproof plastic eventually marketed as Plexiglas.

that DuPont got the American rights, but he was still nonplussed by the lack of cooperation.

In 1938 Haas visited Ludwigshafen again, impressed how the military buildup and the drive for self-sufficiency were fueling growth. South toward Oppau, Haas saw "a large number of new units . . . the latest in chemical factory construction." Man-made rubber was one innovation that struck Haas as a good specialty for his company; in fact, his labs had just started researching acrylonitrile, realizing it might be used to make rubber. His inquiries about American rights were directed to ter Meer, who would not commit to licensing the technology.[35]

As the war drew near, I.G. scientists became more circumspect, less willing to share technical data. By mutual agreement the Philadelphia leather lab sent the I.G. quarterly reports, detailing how American tanneries responded to BASF products like Tamol and Leukanol. The acrylic contract also stipulated the exchange of research data on polymers. But when Harry Neher, chief chemist at Bridesburg, sent Walter Reppe, head of the main laboratory at Ludwigshafen, some questions, the German hesitated. "It seems that several times lately they have been criticized by the government for letting information go out," Haas noted. Reppe advised Haas to avoid the red tape by sending scientists over to ask questions face-to-face.[36]

Haas based his relationships in Germany on honesty and commitment. Erwin R. Sauter provides one good example. In May 1933 Otto Haas hired Sauter, a chemist who had studied under Hermann Staudinger at the University of Freiburg's Chemical Institute, to work on lubricating oils. An expert on X-ray crystallography, Sauter had published on organic compounds like rubber and fiber. Whether in North America or Europe, laboratories fought over experts like Sauter. Sauter did not stay in Philadelphia for long. Homesick for his native land, in early 1934 he returned to Freiburg, but once there he continued to do research for Rohm and Haas under Staudinger's direction.[37]

A similar mutual respect characterized relations with Kurt Albert and Theodor Goldschmidt—and for many years with BASF and the parent I.G. Farben. By 1938, however, caution and fear were beginning to interrupt the flow of information and fray the bonds of trust of every American connection to Germany. In the case of I.G. Farben rapid growth—"Ludwigshafen is under great pressure"—bred inattentiveness, while also slipups spoke to larger troubles. In a session at the I.G. Farben Building on the syntan contract Haas was confounded by two directors who seemed preoccupied and poorly informed. In talks about alkyl phenols Ludwigshafen scientists were circumspect when Haas pointed out that the I.G. might infringe on a Rohm and Haas patent. "They admitted that this is correct—as much as I.G. admits anything."[38] Between the distraction and the reticence Haas could see that this was not the firm that licensed Ordoval back in 1921. Something had changed dramatically—and not for the better.

The Complexity of It All

In late March 1937, Edgar Kirsopp sat opposite Alfred F. Lichtenstein, president of Ciba Company, the New York branch of the Swiss dyestuffs giant. For the past few years Rohm and Haas and Resinous Products had used Ciba patents in seeking to develop new varnishes, lacquers, and molding powders—but the research had not yet produced a successful product. Nonetheless,

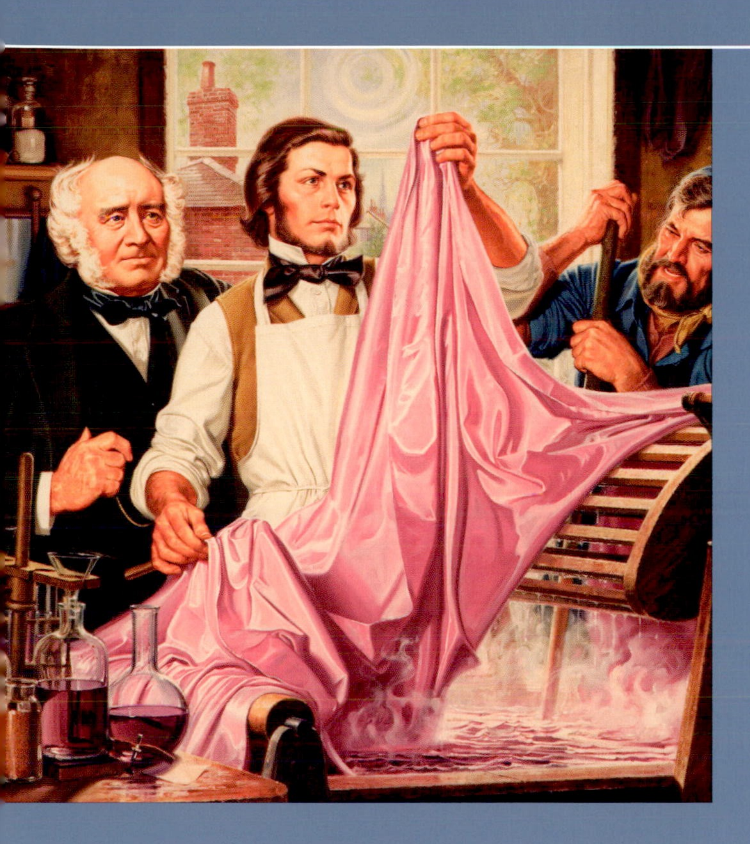

Rohm and Haas started manufacturing chemicals to affix vat dyes after American textile manufacturers were cut off from their European dye and chemical suppliers during World War I. The field of synthetic dyes was pioneered by British chemist William Henry Perkin in 1856, when Perkin successfully synthesized the color mauve. Perkin is shown in a detail of a twentieth-century painting, above (center), examining a test dyeing of silk taffeta with his famous mauve aniline dye.

Fashion's Wheels Turn . . . with Chemistry's Help

A revolution in fashion was launched in 1856 when a young British chemist, William Henry Perkin, figured out how to synthesize the color mauve—a rare dye that came from mollusks—from common coal. Perkin's breakthrough marked the beginnings of the synthetic organic chemical industry—and the ceaseless quest to create more and better textile colors. A half century later the French chemist Hilaire de Chardonnet invented rayon, a man-made silk substitute that became a favorite for ladies' dresses. These developments led to a synthetic textile era that in turn created a market for complex chemicals used in fiber processing and dyeing. Many of these materials were supplied by Europe's advanced chemical industry—until World War I.

The American textile industry found it progressively more difficult to secure supplies of European dyes and chemicals, particularly after the United States entered the war in April 1917. Philadelphia was the nation's largest textile center, and the interruption of transnational trade did not escape the notice of Otto Haas. Rohm and Haas started making sodium hydrosulfite and sulfoxylate formaldehydes, chemicals used to affix the new vat dyes to textiles. The company's sodium hydrosulfite, marketed as Lykopon, soon became the standard specified by American dyers.

The Rohm and Haas tradition of sales plus service established trust in the dyeing, printing, and finishing sectors and opened the doors to collaboration. The firm learned about these new customers bit by bit, feeling its way through territory that, much like leather, was subject to shifts in style and taste. "Formopon is used for calico printing," Haas wrote in one annual report. "Due to a change in the fashions, the printing mills have been rather slow in 1924." Two years later sales of Formopon Extra and Protolin, two products used to remove color from recycled wool, were in decline "owing to the fact that light shades have been popular for some years." Perkin's discovery had a lasting effect, creating the possibility for sudden, unpredictable changes and the need for the manufacturer to stay close to the customer.

Rohm and Haas's research laboratories helped the company steer through this uncertain market by inventing whatever textile chemicals were needed. Products that contributed to the bottom line included the Tritons, a line of synthetic wetting agents and detergents, and the Rhonite and Rhoplex resins for textile finishing. The Rhoplexes were used to make crush-resistant velvet, washable glazed chintz for upholstery and draperies, and wrinkle-proof linen that gave men's suits a crisp, professional look in the dog days of summer. By 1940 the textile department was the firm's most successful division, accounting for over almost one-third of the $2 million plus in profits.

Lichtenstein insisted on adjusting the Rohm and Haas royalty rate, laying a complicated new fee schedule on the table. Known for his orderly legal mind, Kirsopp was exasperated. "The system," he later wrote, "was altogether too complex."[39]

If the Ciba deal made Kirsopp's head spin, there was nothing he could do about it. Despite its investment in U.S. labs, Rohm and Haas was still shackled to European and supremely to German technology, expertise, and political machinations. Otto Haas tried to break free, in part by signing deals with American companies to share research and build export markets. With Atlantic Refining, a Philadelphia oil company, the plan was to develop a viscosity index improver for lubricating oils and to expand sales in Latin America and sub-Saharan Africa. Through this venture Haas introduced new products like Lethane, an active ingredient in insecticides, to Brazil, Uruguay, and South Africa, building on his own limited overseas sales in the prewar era.[40] But for the most part Rohm and Haas and its ancillary firms, Lennig and Resinous, sold products derived from German technology in the United States and Canada.

Rohm and Haas faced major obstacles in absorbing and learning about German technology and then adapting it to U.S. markets, as did all American chemical companies.[41] The Tego experience showed how a feisty battle to win the American rights to a product could not by itself ensure success. Forging deals with I.G. Farben on hydrosulfites, syntans, and acrylics required patience and diplomacy. By the late 1930s this relationship demanded constant attentiveness, with one eye to the Frankfurt executives who counted the royalties and the other to the U.S. Justice Department watchdogs who reinvigorated the antitrust laws during the later part of the New Deal. The good news was that Rohm and Haas—like much of the American chemical industry—continued to grow throughout the challenging years of the Depression.

Nothing was black and white, either in Germany or in the United States. The web of connections grew more intricate, even as World War II approached. In 1939 Haas was pleasantly surprised by BASF syntan sales, which had been steadily rising and surged that year, thanks to a fad for white leather shoes at the high end of the market.[42] As the European war of late 1939 morphed into the global engagements of 1940, Haas noted how "there were heavy demands on the United States and Canadian manufacturers to produce so-called aircraft plywood, mainly for the British Government. . . . This type of plywood can only be satisfactorily produced with Tego."[43] Closer to home the U.S. military created a market for another German invention, acrylic plastics, as the army and navy rapidly built their own fleets of warplanes. Soon Philadelphia-made Plexiglas would become the company's best-known product and a vital ingredient in the growing U.S. aerial war against Germany. Neither the irony nor the complexity was lost on Otto Haas, who contemplated it all from his office in Washington Square.

Otto Haas rarely posed for the camera and always wore a hat outdoors. This candid shot of him leaving the Washington Square office, left, was surreptitiously snapped by a photographer who was taking pictures of the building. Below is a site badge from the Washington Square facility. Opposite: Resinous Products not only shared offices with Rohm and Haas at Washington Square but also had research and production facilities at Bridesburg, including this laboratory, circa 1930s.

03 PLEXIGLAS TRIUMPHANT

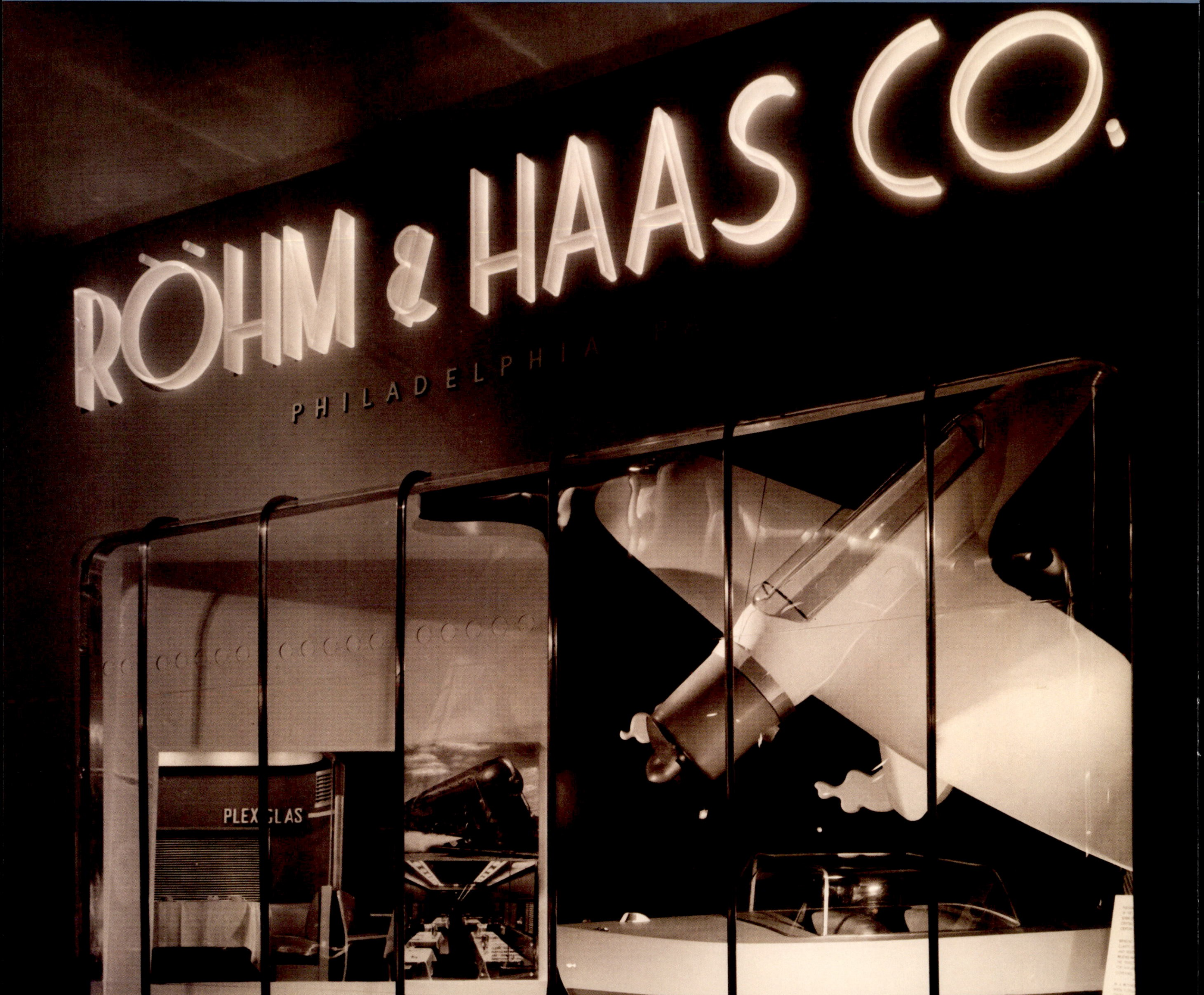

RÖHM & HAAS CO.
PHILADELPHIA
PLEX GLAS

The World of Tomorrow

In April 1939 the New York World's Fair put the World of Tomorrow on display. Dozens of corporations contributed to an extravaganza that celebrated the brave new world of science and technology. Depression-weary consumers ogled early electronics like the RCA television; a Westinghouse robot named Elektro the Moto-Man; nylon, the world's first synthetic fiber; and remarkable plastics like Lucite and Plexiglas. The plastics age had dawned, and innovators like DuPont, Eastman Kodak, and Rohm and Haas seized this opportunity to show off.[1] Rohm and Haas used the World's Fair to advertise Plexiglas, its new acrylic plastic. Plexiglas was a clear, glasslike material that had extraordinary physical and optical qualities. Shatterproof, resilient, and lightweight, it could be cast into large sheets and then re-formed into curved sections. These properties were unprecedented. Among its

AT THE 1939 NEW YORK WORLD'S FAIR THE ROHM AND HAAS EXHIBIT, A THOUSAND-SQUARE-FOOT DISPLAY, ADVERTISED THE VERSATILITY AND MODERNITY OF PLEXIGLAS. THIS LARGE CASE, FRAMED IN THE ACRYLIC PLASTIC, SHOWED APPLICATIONS FOR AIRPLANE CANOPIES, BOAT WINDSHIELDS, AND THE INTERIOR OF HENRY DREYFUSS'S ART-DECO TRAIN, THE 20TH CENTURY LIMITED, CREATED IN 1938 FOR THE NEW YORK CENTRAL RAILROAD SYSTEM. PREVIOUS SPREAD: DURING WORLD WAR II, PLEXIGLAS WAS THE BEST-SELLING PRODUCT AT ROHM AND HAAS. THE COMPANY MADE THE ACRYLIC SHEETS AND FABRICATED THEM INTO COMPONENTS FOR MILITARY AIRCRAFT. HERE, WORKERS AT THE BRISTOL PLANT MACHINE AN AIRPLANE CANOPY.

The Rohm and Haas exhibit at the 1939 New York World's Fair included push-button displays that demonstrated the technical capabilities of Plexiglas and Crystalite. Designer Gilbert Rohde happily reported that fairgoers marveled over the seven mechanical shadowboxes, which showed how acrylics transmitted light, were bendable, and were light in weight. In the center of the room, right, is Alexander Calder's stabile, winner of the Plexiglas sculpture contest.

many uses, Plexiglas was an ideal substitute for glass in airplane windows. In 1939 its potential still lay in the future.[2]

The Rohm and Haas exhibit in the Hall of Industrial Science, Chemicals, and Plastics touted Plexiglas as a technological and aesthetic marvel. Push-button displays showed how this revolutionary plastic transmitted colored light around bends without diffusing it to the sides, a quality that appealed to nightclub owners and jukebox designers, as well as aircraft engineers. Alexander Calder's Plexiglas stabile, the winner of a sculpture competition sponsored by Rohm and Haas and the Museum of Modern Art, pointed to the material's artistic potential. A Plexiglas flute and violin, crafted in Darmstadt, demonstrated its suitability for musical instruments.[3]

Everywhere that fairgoers went, they ran into Plexiglas. Just within the main gates the Highways and Horizons Building of General Motors (GM) beckoned visitors with the world's first illuminated Plexiglas sign, spelling "General Motors" in translucent white letters. Inside the GM building crowds flocked to the "Glass Car," a see-through Plymouth with a streamlined body crafted in Plexiglas. The "X-Ray" sedan's mechanical features—the Deluxe 6 engine, a Hypoid rear axle, Duflex rear springs, and the safety gearshift—were exposed for all to see. Consumers found more displays of acrylic plastic throughout the fair. The Public Health Building had an enormous Plexiglas man, with all his organs visible. The Home Center featured a full-scale Plexiglas house, showcasing Plexiglas interior decorations. As Donald S. Frederick, a plastics expert at Rohm and Haas, recalled, "The World's Fair in New York was something of a boost to us. . . . The General Motors sign on their big building was Plexiglas and we saw the possibilities."[4]

As American tourists marveled at Plexiglas in the World of Tomorrow, dark storm clouds were gathering over Europe. In September 1939, Hitler attacked Poland. In response Britain and France declared war on Germany. Americans rightly suspected that U.S. neutrality might not last. Before long the fair's theme of a flawless future would make Americans uneasy and would seem jejune. The dream of a technological utopia went on hold, as miracle materials like Plexiglas were called to arms.

Plexigum Roller Coaster

Otto Haas first glimpsed the wondrous world of plastics in 1928 when Otto Röhm told him about an intriguing discovery in the Darmstadt labs. Tinkering with applications for acrylic resins, Röhm's chemists had found a way to make automotive safety glass from a rubbery adhesive called Plexigum. When sandwiched between two pieces of flat glass, Plexigum transformed the assemblage into a transparent whole that did not shatter or splinter.[5]

The earliest plastics—celluloid, introduced in 1869, and Bakelite, invented in 1907—imitated natural materials like ivory, tortoiseshell, shellac, and rubber. Plexigum was one of several other synthetic polymers that chemists conjured from out of their test tubes in the 1920s, marking the dawn of the modern plastics age. The new plastics included cast phenolic resins, urea-formaldehyde resins, and polyvinyl chloride, all with unique characteristics. Designers enthusiastically embraced them, introducing the costume jewelry, doorknobs, and radio cabinets that gave art deco its distinctive, colorful look. Acrylics, a family of plastics with good weather resistance and brilliant translucence, were one favored grouping. Plexigum itself was the first step in this direction.[6]

The excitement over plastics coincided with shifts in the auto market. No longer a novelty, the "automobile for everyman" had to be a well-designed vehicle. Jazz Age consumers were growing tired of the no-frills Model T. Comfortable vehicles with hard roofs, real windows, colorful paint jobs, headlights, and other amenities were becoming more popular. In one of the most famous battles in American business history, GM usurped Ford's leadership position through trade-ins, installment credit, stylish designs, the annual model change, and the car for "every purse and purpose." Engineering took a back seat to marketing.[7] Closed cars sporting glass windows, however, created a deadly liability that needed to be addressed.

Fatalities and injuries from broken windshields pushed Detroit to find alternatives to conventional glass. In an accident glass would break, splinter, and fly into the car. In 1927 Ford became the first American automaker to outfit vehicles with a patented safety glass, Triplex, made by sandwiching celluloid film between two sheets of glass. The celluloid adhesive prevented broken glass from flying around. Other automakers followed suit, but early safety glass was far from perfect. Although shatterproof, it turned yellow with long exposure to sunlight, limiting the drivers' visibility and creating a new danger.[8]

Haas suspected that Röhm had stumbled on a gold mine with the crystal-clear Plexigum. The eureka moment occurred in 1927 when lab workers, casting experimental ceiling tiles, pressed some acrylic polymer between two glass plates and applied heat. After cooling, "the glass plates could no longer be separated from one another," wrote lab researcher Adolph Gerlach, "and were also firmly bonded to the polyacrylic layer." When a nail was hammered into the plates, the glass shattered but stayed in place. So was born Rohm and Haas's first plastic product, a new type of laminated safety glass. As Darmstadt's patent applications circulated through the Weimar bureaucracy, the two Ottos corresponded about the discovery. In March 1928 the German firm patented its Plexigum-bonded glass as Luglas, named after one researcher's wife Luise. With the Deutsches Reichspatents in hand the two entrepreneurs brainstormed on how to introduce the product in the United States.[9]

The American Window Glass Company, a Pittsburgh corporation that dominated the flat-glass industry, seemed like a suitable collaborator for the new safety-glass venture. It produced drawn sheet, a type of

General Motors' Highways and Horizons Building at the 1939 World's Fair, above top, was emblazoned with a spectacular white Plexiglas sign that lit up at night. A favorite attraction in the General Motors building was the transparent Plexiglas Plymouth by the Fisher Body Division. The 1939 car was so popular that Fisher created a new "X-Ray" Plymouth for 1940, shown above bottom in a souvenir postcard.

The Machine Age

Gilbert Rohde (1894–1944) was a pioneer in industrial design, a profession that flowered during the Great Depression by giving old products a new look. After studying at New York's Grand Central School of Art and the Art Students League, he devoted his early career to advertising illustration and department-store display. Enthralled by surrealist paintings exhibited in New York and by the European decorative arts he saw at international fairs in Berlin and Paris, and intrigued by new trends in technology, Rohde turned to furniture design. His signature style combined traditional materials like wood with Machine Age novelties like plastics. By the late 1930s Rohde was well known for his modernist furniture for Herman Miller and as founding director of the Design Laboratory, modeled after the Bauhaus.

Rohde became connected with Rohm and Haas through plastics impresario Donald S. Frederick, who needed someone to design the Plexiglas exhibit at the New York World's Fair. A recommendation from Robert D. Kohn, one of the fair's architect-planners, convinced Frederick that Rohde was right for the job. "He is, in my opinion," Kohn wrote, "an excellent designer, and what is more, he is an imaginative one."

The Competition for Sculpture in Plexiglas, jointly sponsored by the Museum of Modern Art (MoMA) and Rohm and Haas, was Rohde's idea. Contestants were encouraged to showcase Plexiglas's unique qualities, including its bendability and translucency. Three discerning aesthetes—art patron Katherine Sophie Dreier, French-American sculptor Robert Laurent, and MoMA curator James Johnson Sweeney—evaluated the 250 entries and awarded the first prize of $800 and a prominent place in the Rohm and Haas exhibit at the fair to Alexander Calder. "This sculpture is equally interesting and strong when seen from any side," MoMA announced in May 1939, praising Calder's creative use of Plexiglas rods and sheets and electric illumination. "The manner in which this sculpture exploits these properties makes light an organic part of the design."

Rohde helped promote Rohm and Haas in other ways. His Plexiglas designs for a coffee table, a chair framed in stainless steel, and an educational display at Rockefeller Center caught the attention of the *Christian Science Monitor*, which publicized them in an October 1939 article, "Furnishings in a New Material." However, the best example of Rodhe's work for Rohm and Haas was the exhibit at the New York World's Fair, a tribute to Machine Age design and a tour-de-force in Crystalite and Plexiglas.

Industrial designer Gilbert Rohde's passion for modern materials led to imaginative designs in Plexiglas. The coffee table in this 1939 advertising photograph for the Valley Upholstery Company, above, has a plate-glass top and a serpentine Plexiglas base, showing one of the first applications of colored Plexiglas. The collaboration between Rohde and Rohm and Haas might have resulted in additional consumer goods if not for World War II. Plans to develop a line of Plexiglas consumer products were curtailed when national-defense priorities channeled Plexiglas and Crystalite into the war effort.

Haas mediated between Darmstadt and Pittsburgh, even as he also envisioned the growing research capabilities of his own chemists and authorized the Bridesburg and Bristol labs to experiment with Plexigum. However, problems plagued the project from the start. . . . While Plexigum had great promise, it appeared to have even greater technical hurdles.

glass compatible with a gummy adhesive like Plexigum. In April 1928 Haas gave Luglas samples to American Window Glass's president, his friend William L. Monro, who was "very much interested in a good binder." When factory trials looked promising, Haas pressed Röhm: "Would it be possible to make the material here?"[10]

Haas mediated between Darmstadt and Pittsburgh, even as he also envisioned the growing research capabilities of his own chemists and authorized the Bridesburg and Bristol labs to experiment with Plexigum. However, problems plagued the project from the start. For instance, in December 1928 Haas conveyed Monro's concerns to Röhm: "The American Window Glass Company thinks there is a big field for Plexigum if you can make it dry quicker so that the time of pressing can be shortened." While Plexigum had great promise, it appeared to have even greater technical hurdles.[11]

Back in Darmstadt, Röhm contracted with glass companies to make Luglas for safety goggles, gas-mask lenses, and windshields, which he exported to distant markets like India and Sumatra. The firm's advertisements touted the Luglas in Reichspresident Paul von Hindenburg's Mercedes-Benz. A grateful father in his turn lauded Luglas: "My son was hurled head-first against the windshield and only has the shatterproof Luglas to thank that he didn't sustain injuries from the glass fragments." Awestruck bystanders gathered bits of the broken windows, souvenirs of the miraculous, new synthetic age.[12]

The struggles with Luglas came at an uncomfortable moment for smaller chemical companies. In Europe and America protectionism led to rising prices for crucial raw materials. Tariff wars encouraged large firms to ramp up their R&D. In the United States, DuPont started the research that would generate neoprene and nylon. A major British consolidation channeled resources into Imperial Chemical Industries. In Germany six major chemical companies formed I.G. Farben, a colossus with superior expertise and finances. Small firms struggled. In the 1925–26 German downturn Otto Röhm in Darmstadt cut his budget and let workers go. Meanwhile, the I.G. launched its own research on acrylic compounds and by 1930 held a Reichspatent for the polymerization process.[13]

By the late 1930s Otto Röhm's company in Darmstadt was using the Plexiglas name to market a wide variety of acrylic products. This 1939 advertising brochure, above, shows young women admiring a set of Plexiglas salad servers.

While the I.G. flexed its muscles, Röhm fretted over his account books. He had brilliant polymer chemists, but his plastics division was not making money. Oropon and other leather compounds barely kept Darmstadt afloat. Bankers and tax collectors pounded on his door. In 1929 the stock market crash on Wall Street portended a global economic crisis. In Philadelphia, Haas wrote: "Darmstadt is in financial difficulties; they have invested about one and one-half million marks in the Plexigum venture and the end is not yet in sight. They have no money to continue; but if they stop now, the investment is lost. There is no doubt that the product has merits;

EXACTING STANDARDS

Attention to quality ensures that Rohm and Haas's customers receive the quality specialty materials they require to make their own products perform as expected. Here, plant operator Lloyd Wood monitors and records temperature readings on a batch of emulsion.

ROHM AND HAAS COMPANY | **TODAY**

Plexite's soundproofing ability was vaunted in the 1936 brochure above, which advertised the acrylic plastic's use as a glazing material for airplanes, broadcasting and telephone booths, homes, hotels, factory offices, hospitals, and trains. The style of artwork, modern for that time, was used to show Plexite as a high-tech material.

but a good deal more work has to be done." The friendship proved its resilience as Otto Haas agreed to underwrite Darmstadt's plastics laboratory. With U.S. dollars Röhm could expand Plexigum research and develop other applications. Haas also ramped up the Bristol effort to improve the Plexigum manufacturing process, reporting that "a development of our organic laboratory which shows promise is the use of acrylic acid esters for laminating glass, to take the place of nitrocellulose films. This problem came to us from our Darmstadt friends." If Röhm's research were to yield products marketable in the United States, Haas would own the American rights.[14]

In 1932 the project stalled, along with the world economy. Haas suspended his Darmstadt subsidies, only to receive woeful letters from Röhm. Haas responded with a vivid description of the devastation of the U.S. Depression. "The business situation here is getting very serious. The industries which we supply are going through one of the worst crises, and the end is not yet in sight. . . . The prices in the leather and textile industries have gone to pieces, with the result that everybody is caught with big inventories. In their trouble, our customers ask for lower prices." In desperate straits Röhm contemplated a sellout to the I.G., but Haas implored him to hang on. Aware that he had the American rights to Plexigum, Haas reinstated his research subsidy.[15]

Haas focused his energies on renewed cooperation with the American Window Glass Company. His production people assisted in the design of a continuous process for laminating the glass; his sales staff helped explain Plexite—the American name for Luglas—to automakers. In 1934 Haas opened a Plexigum production line at Bristol, only to receive alarming reports from American's Monro. The adhesive bubbled when heated, and the windows were not easy to install in cars. If this were not enough, automakers demanded lower prices. It had become nearly impossible to sell anything to GM and Chrysler. Libbey-Owens-Ford had a long-term contract to make GM's safety glass; Pittsburgh Plate Glass had a similar arrangement with Chrysler. Plexite's first major customer, Ford, had suffered dramatic losses. When Monro pressed Ford for more orders—Plexite rear lights were in half its trucks—Edsel Ford dug in his heels.[16]

Salvation came from a different, if familiar, quarter: Röhm's abilities in research. As a chemist fascinated by the whole acrylic-acid family, Röhm had continued his experiments on various compounds, including methyl methacrylate, or MMA. In 1933 Darmstadt chemists created plastic panels by submerging MMA in a polymerizing solution between two pieces of silica glass. By that summer Röhm had registered the trademark Plexiglas in Germany. Fascinated with the material's transparency, brilliance, and malleability, Darmstadt chemists formed it into dinner plates, pipes, and a flute "said to have a much better tone than wooden flutes." Röhm outfitted his own car and the factory's trucks with Plexiglas side windows. By August 1936 the multiple problems of "scaling up" to commercial production were under control, and Röhm proudly unveiled Plexiglas to his professional colleagues, the General Assembly of German Chemists, in Munich.[17]

Bringing Plexiglas to Market

In Philadelphia, Otto Haas was well versed in these new realities, and he prepared accordingly. September 1936 saw Don Frederick on the long train ride from Pennsylvania to California. The vivacious young salesman was to visit airplane companies on the West Coast. Good news had already arrived from the Army Air Corps at Wright Field, which tested high-tech materials for military planes. "Tests on Plexiglas," wrote Lieutenant Colonel O. P. Echols on September 2, led army aviation experts to recommend "its use in cockpit enclosures."[18] Frederick itched to meet aircraft engineers face-to-face and tell them about Plexiglas.

Don Frederick had joined Rohm and Haas in June 1934 with a doctorate from the University of Illinois, having studied under Carl S. Marvel, a leader in polymer chemistry. He started work in Harry Neher's lab at Bristol, testing acrylics, using data from Darmstadt. In January 1936 Haas sent Frederick, who had also attended the University of Munich and spoke fluent German, to Darmstadt to study the Plexiglas setup. Frederick met Röhm and his key science people; saw the factory synthesize MMA; and watched workers cut, bend, and shape Plexiglas. He returned with firsthand knowledge about the new material to share with the Bristol semiworks.[19]

Frederick soon moved from Bristol to Philadelphia and from research to sales. With active rearmament and the threat of war in Europe, the U.S. Army and Navy pressed Congress to expand the air forces, lobbying for authority to build more than 1,100 planes in 1936–37.[20] Army and navy engineers sought to design a fleet from high-performance materials, working at Wright Field in Dayton and the Naval Aircraft Factory in Philadelphia. The search was on for synthetic materials with superior physical characteristics: glass windows had to go. Civilian air carriers had long reported that cockpits outfitted with glass could not withstand collisions with birds. Early planes flew at low altitudes, moving at 180 miles per hour, while ducks migrating at the same level flew at about 60 miles per hour. When a plane hit a duck flying toward it, the impact released twenty times the energy of a .45-caliber bullet striking an armored truck.[21] "Last Saturday one of our airliners had a four-pound duck go through the windshield on the co-pilot's side," explained an executive. As the window shattered, shards, dust, and feathers scattered around the cockpit. "The difficulty of handling the ship was increased and aggravated by a blast of air coming in through the broken window."[22] Military aircraft also slammed into birds. In one instance a duck crashed into an army plane's windshield, striking "the pilot with such force as to cause him to lose consciousness momentarily."[23]

The need for clear, shatterproof glazing led military engineers to plastics. The first glass substitute, cellulose nitrate, could not tolerate the sun. After a few months windows crazed and clouded up. Cellulose acetate had greater durability but could crack and break. When Frederick introduced Plexiglas to the Army Air Corps in early 1936, no plastic yet filled aviation's needs for transparency, nonflammability, and impact resistance. Plexiglas did not discolor, scratch, or break under

Don Frederick (far left) was chosen by Otto Haas (second from left) to lead the sales efforts that made Plexiglas an invaluable material during World War II. Also pictured above are Ralph Connor (far right) who, like Frederick, would go on to become a vice president of the company, and (second from right) an unidentified man. During World War II, Frederick's sales department began publishing the Plexiglas Fabricating Manual, left, to educate aircraft companies and army-navy airfield personnel about the care and handling of acrylic plastics. The company continued to issue instructional booklets well into the postwar era, when they were used by sign fabricators, novelty manufacturers, and hobbyists who made Plexiglas furniture.

The Douglas B-19 bomber—which in 1942 was the largest airplane ever built, at eighty-two tons—was glazed with Plexiglas, below. Opposite: Polishing and buffing were the final steps in the production of bomber noses. At the Bristol plant a war worker wipes down the canopy with a cloth before it is to be polished with power buffers.

sudden temperature changes: it quickly became the army's preferred airplane glazing material.[24]

Despite smarting from his Plexite experience, Otto Haas courageously built a Plexiglas plant in Bristol in 1936, followed by a Crystalite facility in 1937. (Crystalite, a sister product to Plexiglas, was a colorless acrylic powder that plastic molders used for clock faces, speedometer covers, and novelties.) "We are hard at work to find markets for these products," Haas optimistically told stockholders, "and it is a pleasure to say that they have aroused an unusually wide interest, which we hope we can translate into sales in the course of time." Haas could not predict the future, but he clearly recognized the potential. "We feel that the development of the acrylic resins is one of our most important problems," he said in 1937. "It promises, if handled right, to be a big development." By 1939, in a buoyant but philosophical mood, Haas observed how "developments in the plastic field are very swift, almost violent; their exploitation requires a large outlay of capital, first-class technical skill and the handling of a large number of workers. Furthermore, in order to survive, we shall need to have access to the basic raw materials. This is a new situation for us."[25]

Thanks to the courage of his leader, Don Frederick had some highly sophisticated resources at his disposal. The acrylate laboratory in Bridesburg, expanded in 1937 and directed by Harry Neher, focused on basic polymer research but shared those discoveries with Bristol, which evolved into plastics central. The Bristol plastics laboratory, under W. S. Johnson, collaborated with Neher's group in the development of sheets, rods, and molding powder. The Bristol physics lab, directed by W. F. Bartoe, resolved queries about the plastics' physical properties. Collectively, the labs did research on acrylics, created salesman's samples, and worked on customers' problems. Bristol scientists also assisted foremen with manufacturing problems in the plants that polymerized MMA, made Crystalite powder, cast Plexiglas sheets, and eventually formed canopies for fighter jets and windows for naval destroyers.[26]

Frederick and a small sales staff marketed Plexiglas and Crystalite to a variety of customers. The plastics industry consisted of chemical companies that made powders, film, and sheeting; small shops that extruded, molded, or shaped consumer products; and firms that molded and installed plastic parts into radios, telephones, cars, ships, and planes. Just as Otto Haas did with Oropon, Frederick had to educate customers, from big businesses to small family-owned mold shops, on the advantages of acrylic plastics. Sometimes companies saw their competitors succeed with Plexiglas, making his job easier. When the Unbreakable Lens Company put Plexiglas lenses on the market, the American Optical Company took notice. In other cases customers had little use for Plexiglas or wanted it for oddball applications. General Electric engineers did not think Plexiglas had a future in lighting devices, but they asked whether it would make good refrigerator trays and doors.[27]

Eight Miles High

Indicative of changing times was Don Frederick's 1937 report that "the Douglas Aircraft Company has just been awarded a contract by the U.S. Army for 177 airplanes. These planes are entirely equipped with Plexiglas." Frederick also received a tantalizing letter from Wright Field. If the army were to make high-flying planes, asked J. B. Johnson, chief of the materiel division, would Plexiglas work in the cockpits? Frederick consulted Bristol's physics lab: was Plexiglas strong enough to tolerate atmospheric pressure at 34,000 feet? With Bartoe's report in hand Frederick excitedly told Johnson, "Tests seem to show that Plexiglas is a very satisfactory cabin window material

ROHM AND HAAS COMPANY | **TODAY**

OPPORTUNITIES FOR ENRICHMENT

Rohm and Haas offers employees continuing education opportunities that keep them informed of the latest technology and managerial advances. Here, Joanne Sekella, Sheila Van Wicklen, Lawrence Cheam, and Christine Millaway attend a "Train the Trainer" course.

After World War II, Plexiglas was given a new military role in transparent boards used for aircraft tracking. The U.S. Air Force, formed in 1947, transferred its Radar School from Boca Raton, Florida, to Keesler Field in Biloxi, Mississippi. Behind the transparent acrylic plastic sheets, right, Keesler personnel receive reports from radar operators and chart the changing positions of aircraft, writing on the status boards in reverse with special crayon on the back surface of the Plexiglas. Everything is illuminated by light traveling through sources concealed at all edges.

for stratospheric airplanes." The main challenge would be finding the right cement. The lab buckled down to work, taking nearly a year to develop an adequate adhesive.[28]

Aircraft companies had trouble shaping the all-Plexiglas enclosures that the military wanted. Besides high-altitude flying the introduction of aerial surveillance photography and machine-gun blisters created the need for windows with a wider viewing range. The Bristol semiworks sent the Vought airplane factory an all-acrylic prototype, along with directions on how to make it from Plexiglas rods and sheets. For the X0SS-1 seaplane, Boeing Airplane's Stearman division copied the Vought design and sent window prototypes to Bristol for tests. Frustrated by Plexiglas assembly, the Wichita plane maker asked Rohm and Haas to build the windows, but Frederick declined to "tackle the job of furnishing a complete enclosure." The chief plastics officer remained tight-lipped, even as he contemplated the possibilities.[29]

Experiences like these opened Haas's and Frederick's eyes to the profits to be made in fabrication. Some aircraft companies that tried shaping one-piece enclosures themselves did not know enough about plastics. Their problems paralleled installation troubles with Plexite. In July 1938 G. P. Young, one of Frederick's salesmen, visited the Middletown Air Depot in Pennsylvania, where frustrated army fabricators "were attempting to form spherical sections from Plexiglas for installation in the noses of the B-17 Boeing Bombers." When Young described this scene to his bosses, Frederick finally volunteered to build the Plexiglas noses. "We have recently developed a very good method of making spherical sections relatively free from optical distortion," he told Johnson, paving the way for Rohm and Haas to enter the fabrication business.[30] In 1940 Haas opened a fabrication plant in Bristol, where workers formed sheets of Plexiglas into such 3-D enclosures as cockpit canopies, navigator's domes, and power-driven turrets. In early 1941 he opened the South

Gate forming plant outside Los Angeles, where Bristol-made Plexiglas sheets were shaped into enclosures for West Coast airplane manufacturers. By the end of 1941 Plexiglas was triumphant, with its manufacturing and fabricating facilities generating almost half the company's profits.[31]

Plastics at War

When the Army Air Corps gave the thumbs up to Plexiglas in 1936, Rohm and Haas plastics sales were a meager $13,000; in 1937 sales climbed almost tenfold to $119,000. After March 1941, when Congress passed the Lend Lease Act authorizing President Franklin Delano Roosevelt to supply materiel to the Allies, Plexiglas sales really took off. That year Rohm and Haas's plastics sales reached $8.9 million. The United States built more than eighteen thousand military aircraft in 1941 compared with six thousand the year before. Rohm and Haas made 85 percent of the plastics that went into those planes.[32] The company had entered the big time.

Rearmament pushed Haas to redirect resources away from marketing and sales into development and production. Crystalite sales dwindled as Washington restricted the use of plastics for civilian goods. Instead national defense stimulated the economy and created a market for high-tech materials. Just a few years before, Frederick's staff struggled to sell plastics, but now the armed forces and their civilian contractors pounded on the door. "In normal times, it is the Sales Department that gets the lion's share of attention," Haas wrote in 1943. "In war times, the Development and Manufacturing Departments . . . move to the foreground because the most important problems are how to find ways and means to produce quickly what the country needs."[33]

Producing more Plexiglas simultaneously tested and strengthened the firm's abilities in industrial research, engineering, and plant management. Economies of scale allowed the firm to reduce costs and prices. "It was extremely difficult to improve processes and maintain orderly production in all the turmoil caused by rapid expansion," Haas wrote, particularly since a new labor force needed training.[34] However, he credited his plant managers, technical service labs, and research labs with collegial collaboration.

As army and navy engineers designed more sophisticated planes, Rohm and Haas technicians helped aircraft companies with experimental designs of aerodynamic canopies and windows. Interactions among South Gate engineers, aircraft builders, and Bristol scientists show how innovation through collaboration took hold in this new area. South Gate's engineers developed close ties with the Pacific Coast airplane industry, staying "in constant touch with engineering staffs of the different companies" and helping them solve problems. Following in Don Frederick's footsteps,

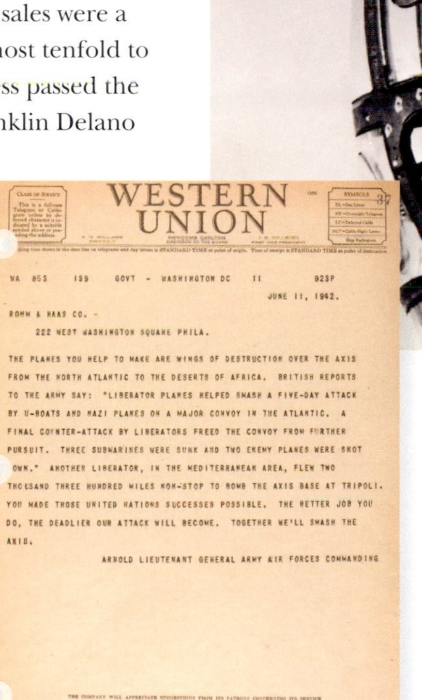

WESTERN UNION

WA 853 139 GOVT - WASHINGTON DC 11 923P

JUNE 11, 1942.

ROHM & HAAS CO. -

222 WEST WASHINGTON SQUARE PHILA.

THE PLANES YOU HELP TO MAKE ARE WINGS OF DESTRUCTION OVER THE AXIS FROM THE NORTH ATLANTIC TO THE DESERTS OF AFRICA. BRITISH REPORTS TO THE ARMY SAY: "LIBERATOR PLANES HELPED SMASH A FIVE-DAY ATTACK BY U-BOATS AND NAZI PLANES ON A MAJOR CONVOY IN THE ATLANTIC. A FINAL COUNTER-ATTACK BY LIBERATORS FREED THE CONVOY FROM FURTHER PURSUIT. THREE SUBMARINES WERE SUNK AND TWO ENEMY PLANES WERE SHOT DOWN." ANOTHER LIBERATOR, IN THE MEDITERRANEAN AREA, FLEW TWO THOUSAND THREE HUNDRED MILES NON-STOP TO BOMB THE AXIS BASE AT TRIPOLI. YOU MADE THOSE UNITED NATIONS SUCCESSES POSSIBLE. THE BETTER JOB YOU DO, THE DEADLIER OUR ATTACK WILL BECOME. TOGETHER WE'LL SMASH THE AXIS.

ARNOLD LIEUTENANT GENERAL ARMY AIR FORCES COMMANDING

The North American B-25 bomber, above, had a rapid-fire machine gun mounted in a panel of a Plexiglas "greenhouse." The bombardier's job was to fight off frontal attacks, while the enclosure permitted all-around vision, a valuable asset on patrol or bombing missions. Left: Rohm and Haas received countless telegrams from military authorities and aircraft manufacturers praising Plexiglas and documenting its role in winning the war.

War on Insects

In 1944 the U.S. Food Administration, a federal wartime agency, asked Americans to grow more fruits and vegetables. Commercially grown food was needed for the troops, and consumers had to become more self-reliant. The year before, Americans tilled twenty-two million victory gardens, producing over 40 percent of the country's vegetables while battling their own set of enemies. Pests like the Colorado potato beetle, the cabbage worm, and the black bean aphid could devastate a victory garden. Rohm and Haas came to the rescue with Lethane 60, a powerful synthetic insecticide.

Otto Haas started his agricultural chemicals business in 1926 at the urging of Charles Hollander, who realized the field was wide open. Two researchers—chemist Charles H. Peet and entomologist Anthony G. Grady—were brought in to focus on insecticides. They identified a class of synthetic organic compounds that were toxic to insects and developed a simple way of killing bugs in a glass chamber and doing a body count. This method—the Peet-Grady test for liquid household insecticides—became an industry-wide practice and in due course was officially adopted by the National Bureau of Standards.

Peet, while searching for a crop insecticide, stumbled across compounds that were effective against the common housefly. They also killed the bloodsucking parasites that caused cattle, sheep, and pigs to drop in weight and decline in value. Rohm and Haas named Peet's discovery Lethane, combining *lethal* and the Greek word for death, *thanatos*. With Lethane's 1929 commercialization Haas transferred Peet from research to sales, in keeping with the company's tradition of having people who understood the science and the market promote the product.

Lethane and other bug sprays were tested at the Bristol insectary, where three to five thousand flies were destroyed each day as part of the testing process. Three young women—Margaret Montalto, Millie Pirollo, and Mary Janico—were in charge of the famous Peet-Grady chamber. Through its portals went the healthiest flies, nurtured to adulthood on a feast of milk and water, and out came insects that were "pretty dead." Montalto described how: "At first, I was squeamish about handling flies, but not anymore. I enjoy this type of work because I know I am doing something that is beneficial to everybody who hates flies—and that includes just about everyone."

Lethane was the first successful product out of the Bristol labs, and the demand for insecticides during World War II helped Rohm and Haas establish a presence in agricultural chemicals. Other important products would follow, including Perthane, a spray used by homemakers against wool-eating moths, and Dithane, a fungicide that prevented mildew on grapevines and accounted for major sales in the vineyards of France and Italy.

Rohm and Haas entered the agricultural chemicals business in 1926 and began to market the insecticide Lethane in 1929. Above, Margaret Montalto is shown counting dead flies twenty-four hours after they were sprayed with insecticide. She had worked at the Bristol insectary for six years when this photo was shot in 1947.

Bristol scientists went out West to lend a hand. "Our chief chemist has called, in company with the South Gate engineers," Haas noted in 1942, "at the more important West Coast aircraft companies to learn their particular problems." As aircraft companies built high-altitude aircraft, they sought reliable physical data on Plexiglas, and Rohm and Haas scientists developed that data and showed aircraft engineers how to use it. Revealing of the prevailing attitude was a July 1942 telegram from the Glenn L. Martin Company: "Many thanks for your fine cooperation in assisting us to complete on time the new nose and tail installations for our experimental airplane."[35]

The same collaborative spirit surrounded the subsequent opening of a Plexiglas plant in Knoxville, Tennessee. The military's need for acrylic plastic led the U.S. Defense Plant Corporation to invest $3.3 million for Rohm and Haas to convert an old auto-parts factory into a Plexiglas plant. After the facility started up in 1943, Bristol chemists, physicists, and engineers advised their southern counterparts on MMA production, sheet casting, and fabrication. The labs in Bristol and Knoxville routinely shared reports on manufacturing problems like smears, pimples, and edge color.[36]

Between January 1937 and July 1942 the output of Plexiglas increased a hundredfold, while prices were reduced from $2.86 per square foot to $1.02. The Army Air Corps praised Plexiglas for improving the vision of the pilot, navigator, and gunner: "The airplane of today has literally been provided with eyes to see." Bristol's 2,400 workers basked in the reflected glory of an Army-Navy Excellence Award in January 1943. Further commendations came in September 1943, October 1944, and May 1945; on each occasion an additional star was attached to the original E Award banner.[37]

As the June 1944 Normandy invasion foreshadowed the European war's end, Otto Haas pondered the firm's future. Since Plexiglas so depended on military needs, Haas anticipated changes after the Allied victory. But he thought the return to normalcy would be gradual, as military orders dwindled and consumer spending revived. Following V-E Day he rescinded contracts and sold the firm's California facility. Even so, Haas was shocked when almost $7 million in war business was cancelled soon after V-J Day.[38] Frederick's sales force staunched the bleeding wound by selling colored Plexiglas sheets to manufacturers of umbrella handles, novelties, furniture, and signs. They also called on factories that made civilian aircraft, cars, musical instruments, refrigerators, and store-display fixtures—revisiting many of Darmstadt's prewar ideas. All this was a far cry from "the eyes of aviation." The final end of the wartime saga came in late 1945, when Haas reluctantly closed the Plexiglas fabrication facility at Bristol.[39]

On July 2, 1943, the Army-Navy "E" Award was presented to the employees and managers of the Resinous Products and Chemical Company in recognition of their aid to the production of plywood, military paints, and strategic chemicals. Otto Haas, wearing a suit coat and tie, stands behind the A and R in the banner, above. Bristol employees compiled a scrapbook of memorabilia from their "E" Award celebration, left, that contained such documents as an award ceremony booklet, photos of the occasion, and congratulatory letters from airplane manufacturers.

After World War II a large market would emerge for Plexiglas in illuminated commercial signage. Right: The popularity of Plexiglas signs among Philadelphia merchants is reflected in this 1979 photograph of 22nd Street stores. This early example of a colored sign from 1940, below, is fittingly a Rohm and Haas advertisement for Plexiglas.

Plastics Redux

During his retirement Don Frederick reflected on his early experience with Plexiglas and the aircraft companies. "They accepted it . . . quite readily," he said, "because it had so many advantages over things they'd been using."[40] But memory can collapse the passage of time and erase the gritty details of innovation. Rohm and Haas established itself as the acrylic leader with Plexiglas, but the path was not as direct as Frederick remembered.

Although Plexiglas swiftly won favor at Wright Field, it took five years for this plastic to generate a profit for Rohm and Haas. This was true for the entire Plexi line. It took more than a decade for Otto Röhm and Otto Haas to abandon their dreams for Plexigum and still longer for Plexite to die a genteel death. Crystalite sales were also slow.

The experimental character of the plastics industry forced customers of Rohm and Haas, who were the actual manufacturers of the end products, to develop patience. By the late 1930s Haas's chemists still had not created a good compression molding powder, as they puzzled over the whole field of injection molding.[41] Darmstadt's eureka moment of 1927, when Adolph Gerlach realized that glass and acrylic could be fused, was followed by years of trial and error. Invention was tough, but innovation was tougher.

Haas steered his firm's acrylic ventures down multiple paths, involving different degrees of risk. He worked within the legal frameworks of the time to secure technical information from his trusted colleague, Otto Röhm. While the two firms shared patents and licenses, their relationship extended far beyond the legal realm, as Haas sent rising stars like Don Frederick to Darmstadt to learn. As the war drew near, the Nazis prohibited German firms from exporting technologies without prior permission, but this did not stop old friends from writing, telegramming, and visiting.

Otto Röhm died on September 17, 1939, heartbroken over his wife's death three years before—she was Jewish and had been persecuted by the anti-Semitic regime—and also sad at the realization he would never see the full results of his pioneering research on acrylics. Haas became the guardian of Röhm's legacy, in a move that blended loyalty and business acumen. He would have pushed out internationally with Plexiglas, but a Justice Department

Plexiglas started life as a good product, but because of the war economy it became a remarkable product. Rearmament brought together parties who would never have collaborated during peace: the chemical company, the aircraft builder, and the military engineer. Together they reshaped the futuristic plastic of the World of Tomorrow into an essential part of aviation.

inquiry, probing the so-called acrylic monopoly between the American defendant, Rohm and Haas, and alleged coconspirators Imperial Chemical Industries, the I.G., and Darmstadt, ended in a consent decree that in effect limited Plexiglas sales overseas.[42]

The American Window Glass arrangement, which began with a handshake between two like-minded businessmen, benefited both Haas and Monro, allowing them to experiment with Plexigum initially free from the constraints of a contract. Their mutual trust and respect comes through in surviving documents, with Haas praising Monro's commitment, integrity, and ability to deal with people like Edsel Ford. Their joint ambition to make Plexigum the laminate of choice for safety glass was never realized, in part because big businesses like Pittsburgh Plate Glass and Union Carbide exerted enormous market power, introducing commodities that undercut specialty materials.

Haas enjoyed his greatest success when the U.S. military stepped into the Plexiglas market, as national security created a new need. Plexiglas started life as a good product, but because of the war economy it became a remarkable product. Rearmament brought together parties who would never have collaborated during peace: the chemical company, the aircraft builder, and the military engineer. Together they reshaped the futuristic plastic of the World of Tomorrow into an essential part of aviation. Don Frederick not only rode this Plexiglas rocket but also managed to steer its course. He spent his entire career at Rohm and Haas, working in an office connected to Otto Haas's by a private door. From Washington Square, Frederick built the Rohm and Haas plastics business, rising from salesman to vice president. When the war was over, the two men together grappled with the challenge of reorienting the potential of Plexiglas.

After World War II other products—chemicals for leather, textiles, and agriculture—contributed to the firm's continued growth along with the international business, but plastics remained the major profit center it had become in wartime. Frederick had developed some customers during the war among the small-scale plastics fabricators that used off-grade material from the aviation industry. Haas believed these customers would become important once the national defense program ended.[43] For once his intuition misled him. The new Plexiglas market lay not in clocks or jewelry but in the burgeoning postwar suburban landscape, where illuminated signs advertised everything from Shell gasoline to the regional mall. Creating those colorful signs involved a new generation of polymer chemists and a new group of collaborators: corporate branding experts, industrial designers, and other image makers. When Don Frederick "saw the possibilities" in the General Motors sign at the 1939 fair, his eyes first glimpsed the postwar future for Plexiglas.

But it was another Don who brought the concept of the World's Fair through the front door of Rohm and Haas. Where others attended the fair to catch a glimpse of the world, Rohm and Haas executive Donald F. Murphy chose to meet it head on, on every habitable continent. It was his interest and dedication that added a new dimension to the Rohm and Haas story, that of innovator to the world.

The advertisement above, which appeared in the 1955 issue of the Reporter, *a magazine for Rohm and Haas customers, shareholders, and employees, suggested some of Plexiglas's colorful possibilities.*

04
THE INTERNATIONAL ARENA

The Man and the Moment

In horn-rimmed glasses and a three-piece flannel suit Donald F. Murphy looked the quintessential organization man. The look was deceiving. In reality Don Murphy loved to travel and experience different cultures. Taking advantage of America's changed, preeminent position after World War II, he combined his interests with a passion for the biological sciences and crafted a successful career, directing Rohm and Haas foreign operations and laying the foundations for the later global business. Murphy was introduced to international markets at their most favorable moment. Europe after World War II presented American corporations with golden opportunities. Between 1947 and 1953 the Marshall Plan's Economic Cooperation Administration helped reconstruct the ravaged European continent, channeling $13 billion to battle-worn Allies. U.S. corporations, unscathed by bombs,

OTTO HAAS TAPPED DON MURPHY TO LEAD ROHM AND HAAS'S INTERNATIONAL EXPANSION IN THE 1940S. HERE, MURPHY (CENTER) TOURS THE SITE OF THE MODIPON SYNTHETIC FIBERS PLANT IN INDIA DURING THE 1960S. THIS COLLABORATION WITH THE MODI BROTHERS (ALSO SHOWN HERE), A LEADING INDIAN TEXTILE FAMILY, WAS ONE OF THE COMPANY'S FIRST JOINT VENTURES. PREVIOUS SPREAD: AGRICULTURAL CHEMICALS HELPED ROHM AND HAAS ESTABLISH A STRONG PRESENCE IN EUROPE AND LATIN AMERICA AFTER WORLD WAR II. HERE, ALBERTO MARTINEZ (LEFT), A MAJOR COLOMBIAN BARLEY PRODUCER, TALKS WITH ALFONSO CHAVARRO (RIGHT), A ROHM AND HAAS TECHNICAL-SERVICE REPRESENTATIVE, ABOUT THE YIELD HE OBTAINED WITH DITHANE.

F. Otto Haas, Otto Haas's older son, above, managed the company from 1960 to 1970 before retiring to pursue philanthropy. Before becoming CEO, F. Otto worked with his father to realize the potential for new business offered by the expanding international economy. As business expanded around the world, the sales department started producing advertising literature in multiple languages, including a brochure in Spanish, right.

were encouraged to share technology, import raw materials, and set up new manufacturing facilities. An entomologist by trade, Murphy had worked on agricultural products during the war. He was ready for fresh challenges when Otto Haas, in his best risk-taking, entrepreneurial style, tapped him to lead the company's international expansion efforts.[1]

Haas gave Murphy the freedom to develop new markets in Europe and to expand existing operations in Canada, England, and Latin America. "As the standard of living rose abroad, there would be a natural market for things that we made here," recalled F. Otto, Haas's older son and his eventual successor as president and CEO. Father and son realized that factories everywhere would "make more collars, make more shoes, make more shirts." For a chemical company "there was a ready market for established products," F. Otto Haas said, "if you just sold them properly."[2]

Between 1945 and the late 1960s Rohm and Haas transformed its modest and stumbling overseas presence into the foundations for a growing global business. Granted a good deal of autonomy, Murphy and his crew targeted Latin America, Europe, and Asia. They established offices, plants, and joint ventures in England, France, Italy, Mexico, India, and Japan. Their exploits—trekking through Brazilian rainforests and across African deserts to get the product to isolated tanneries and farmers—were legendary back at the settled home office in Philadelphia.[3]

Like Haas, Murphy was someone who fostered an entrepreneurial spirit among his staff. Many adventurers who did well in foreign operations moved up the ranks, starting with Murphy himself and two of his protégés, Frederick W. Tetzlaff and Vincent L. Gregory, Jr. After his retirement the international business continued to be a proving ground for leaders like J. Lawrence Wilson, Rajiv L. Gupta, and Pierre Brondeau, who later transformed Rohm and Haas into a truly multinational corporation.

Before World War II

In the 1920s and 1930s Rohm and Haas mostly exported leather chemicals, experiencing mixed results. After closing the Argentine and Chilean offices during the 1921 recession, Haas was slow to expand exports of Oropon elsewhere in Latin America or to places like Australia, Japan, and New Zealand. By the 1930s, however, he admitted that "sales of Oropon in the United States have almost reached the saturation point, but there is a possibility of increasing our Oropon sales in foreign countries."[4] A stronger presence was needed in target markets.

Rohm and Haas adopted the model common among American businesses testing international waters, first selling products through resident agents and then setting up sales offices. In 1936 Haas opened small subsidiaries that imported his products to the United Kingdom, Argentina, and Canada, while also sending his "best technical tannery man, Mr. [K. L.]

Jopke, to revive . . . business in Australia and New Zealand."[5] Two years later Rohm and Haas staffer William P. Rheuby visited Buenos Aires "to advise our representatives, to visit the trade, and to make a general survey of the tanning situation in Argentina, Uruguay, Chile, Peru, and Brazil."[6]

Leather products—bates, syntans, and finishes—were the principal exports, but Haas envisioned a diversified international business. As a European himself and a citizen and resident of the United States, Otto Haas was internationalist by nature as well as by inclination. He pinned high hopes on a London subsidiary, Charles Lennig and Company Ltd., created in 1936 to "take care of the business we hope to develop throughout the British Empire, especially for our insecticides, detergents, etc."[7] George D. Kirsopp, Edgar's son, was appointed the Lennig-U.K. director, briefed to expand the market for Philadelphia exports. World War II curtailed these ambitions.

In 1945 Haas looked abroad once again. "Export sales have never been an appreciable factor in our business," he told stockholders. "A small percentage of our total production is allocated to export sales. In this way, we hope to make some contribution to the rebuilding of world trade, and to laying a foundation for continued export trade when production and demand are again more nearly in balance."[8] The next year he was informing Rohm and Haas stockholders that "there seems to be a good demand for some of our products in Europe, particularly our leather chemicals and plastics. It is expected that our export business will increase in 1947 if we can allocate more materials in foreign sales."[9]

Murphy's Laws

Don Murphy was put in charge of executing this broad vision. The world was Murphy's to conquer, and this he did from positions of increasing responsibility, ultimately retiring as vice president for the international department. Murphy expanded export sales from just under $1 million in 1946 to $144 million in 1969, nearly a third of the firm's total revenue.[10]

Murphy had cut his teeth on agricultural products. In the mid-1920s Otto Haas, curious and entrepreneurial as always, had recognized the potential of organic insecticides. He hired scientist Charles H. Peet to investigate. Peet's research yielded Lethane, the active ingredient in insecticide and cattle sprays commercialized in 1929. In 1930 Murphy, a native Bostonian with a bachelor's degree from the Massachusetts Agricultural College, joined the company. For many years he directed the Insecticide Testing Laboratory at Bristol, which evaluated new compounds—fly sprays, mothproofing agents, ovicides, and fungicides—formulated by the Insecticide Research Laboratory in Bridesburg. He also helped salesmen keep abreast of science and worked with customers on field tests of products. During the war Murphy and others tested Lethane and Dithane, along with new products like the Rhotanes and the Tritons. He was made sales manager for agricultural and sanitary chemicals and, after the war, assistant manager for the plastics and chemicals division. Haas was in part acknowledging Murphy's "exceptional ability as a negotiator with customers, suppliers, and government officials" when he put Murphy in charge of foreign sales.[11]

Western Europe was a ready market for agricultural chemicals like Lethane insecticide and Dithane fungicide. French interest in Dithane became apparent when Pechiney, a leading manufacturer of agricultural chemicals, knocked on the door. Dithane's success in the United States—Florida and Texas farmers had used trial samples to battle an outbreak of potato blight in 1943 and 1944—and tests by a prominent plant pathologist working for the Swiss firm Dr. R. Maag Ltd. pointed to its use on European crops. A protection against mildew on grapes, Dithane

Rohm and Haas's successful insecticide Lethane was promoted worldwide in advertisements such as the one below from a 1958 issue of the trade journal Soap and Chemical Specialties.

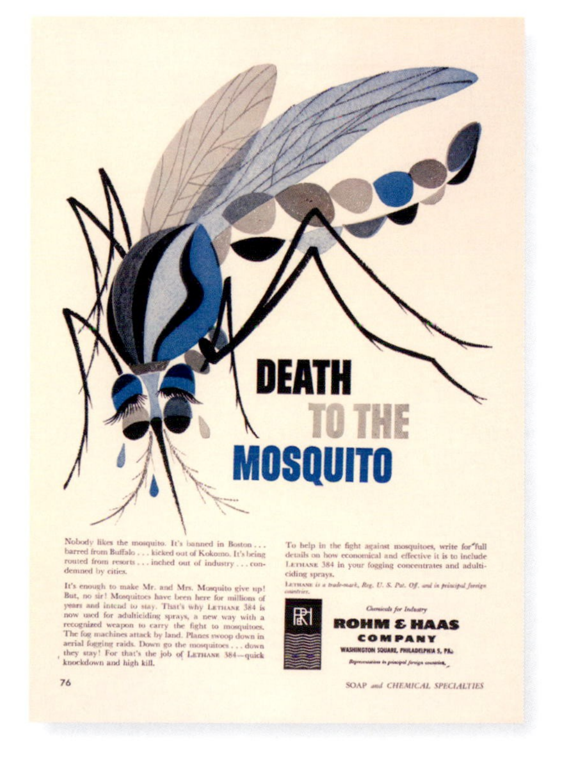

ROHM AND HAAS COMPANY | **TODAY**

NEW AND BETTER USES

Experimental research to find new uses for existing products keeps Rohm and Haas at the leading edge of specialty materials innovation. Here, scientist Patrick Dougherty prepares an experiment using Acrysol SCT-275, a polymer used in paint coatings.

seemed like a good match for vineyards in France, Italy, and Switzerland. Pechiney asked for the French manufacturing rights. "Mr. Moundlic called on us a number of times in Philadelphia," Murphy recalled, "as well as conferring with the Lennig people in London." Needing time to scope out the market, Murphy remained noncommittal.[12]

Murphy's initiation in Europe was a trial by fire. In 1949, in response to Otto Haas's restless entrepreneurial drive, two vice presidents—Don Frederick and Ralph Connor—had brokered a deal to build a chemical plant adjacent to the Shellhaven refinery near London. Royal Dutch Shell was the partner. At the eleventh hour Haas decided to pull out, sending apologetic letters to top executives and to the young international team in London with the bland message that "anti-trust considerations in the U.S. were the main reason for this action." Murphy and Tetzlaff, a Dartmouth M.B.A. who had come to the company from U.S. Rubber, smoothed ruffled feathers and finessed the apology.[13]

Frederick documented what happened next. "On the termination of our negotiations with Shell," he told the Rohm and Haas directors, "we recognized that we must not lose time in arranging for the manufacture of Dithane in Europe, preferably in France." Although Pechiney still pounded on the door, Murphy refused to give any single French firm an exclusive license, believing that multiple distributors could best establish Dithane. Murphy and Tetzlaff traveled throughout France, promoting their message that "several years work in Switzerland have established firmly that Dithane Z-78 . . . is the best available fungicide for the control of scab on apples, pears, and other deciduous fruits, and for rust, leaf spots, and other diseases on ornamentals." Their aim was to find a manufacturing partner, distributors, and approvals from French officials. The timing was right. Disease had hit crops, and harvests were devastated owing to shortages of sulfur and copper sulfate, which were used as fungicides. Agriculturalists needed a good spray. At the Ministry of Industry and Energy the chemical specialist, a Monsieur Pavot, speculated that if France had "another disease year like 1951," the country would "face severe loss in food production." Charles Vesin, the inspector of agriculture in charge of the wineries, confirmed the need for a fungicide for "vines, potatoes, deciduous fruits, and to a lesser extent, vegetables, particularly in North Africa."[14]

Other officials smiled on Dithane. At the American Embassy, Homer Herrmann, the agricultural attaché, and Donald Woolf, a chemical specialist with the commercial attaché, shared their insights on French farming and speculated that "newer materials" from the United States could augment productivity. Maurice Raucourt, the pioneer of phytopharmacy—the study of pesticides and their use on crops—and director of the Station de Phytopharmacie au Centre National de la Recherche Agronomique à Versailles, was satisfied with early French tests on the fungicide and promised "to issue a visa covering the use of Dithane on vines, truck crops, and fruits."[15]

The Ministry of Industry and Energy saw two solutions to the sulfur shortage: Standard Oil's SR 406 or Dithane Z-78. However, massive imports of American goods had drained the French of foreign-exchange reserves, leading trade officials to restrict U.S. imports. Pavot encouraged Murphy to partner with a French firm to build a Dithane production facility in France. The French official promised to remember the American companies that helped French agriculture during this crisis if those firms later sought to expand "operations in the chemical field." The race was on, and Rohm and Haas responded with haste. Murphy and Tetzlaff struck a deal with Prochinor, a chemical works partially owned by the Société de Produits Chimiques et Engrais D'Auby, a large fertilizer company. Established in 1950–51, Prochinor—short for Société de Produits Chimiques Industriels et Organiques—made chemicals from Auby's hydrogen and nitrogen at a facility near Arras in northern France. "It is our impression that Prochinor, like Auby, is up to date, aggressive, and able to move ahead rapidly once decisions are reached."[16]

Rohm and Haas fungicides sold well in the winemaking regions of France and Italy, where they were used to battle diseases like black spot and powdery mildew. Grapes treated with Systhane fungicide, opposite, are harvested at Château Gruaud-Larose, in Bordeaux, France, in 1987. Below: This advertising leaflet for the fungicide Dithane Z-78 was typical of such literature in the postwar era.

Passing the Baton

Rohm and Haas celebrated its golden anniversary in 1959, turning fifty with Otto Haas still in charge. But as 1959 progressed, the "Old Man's" health deteriorated. Undaunted, he worked until the very end. Haas formally announced his retirement to employees on New Year's Eve, and two days later on January 2, 1960, died at his home in Villanova, Pennsylvania.

Otto Haas had been privately planning for his succession since the mid-1940s, when he established several charitable and family trusts and a public foundation (later named the William Penn Foundation), using his share of the company's stock. He also brought his two sons, F. Otto and John, into the firm. After graduating from Amherst College, "Young Otto" obtained a Ph.D. in chemistry from Princeton University, while younger brother John earned a master's degree in chemical engineering from the Massachusetts Institute of Technology. Young Otto worked in the acrylic research lab at Bridesburg and after a stint in the wartime navy went into technical marketing before becoming executive vice president in 1953. The "people-oriented" John spent time at the Bridesburg, Knoxville, and Houston plants before becoming personnel director. In 1960 Young Otto succeeded his father as president and chief executive officer, while the Old Man's confidant and research director, Ralph Connor, was elevated to chairman to provide continued guidance.

F. Otto Haas managed the firm for ten years, helping it through the transition from an entrepreneurial business to a professionally managed company. Young Otto had a very different management style from his father, acting as a team leader rather than a one-man show. He relied on the advice of his senior managers—people like Ralph Connor, Donald S. Frederick, Louis Klein, Donald F. Murphy, and Frederick W. Tetzlaff—who had brought key innovations to market, built the R&D labs, and expanded international sales. He shed the firm of older management techniques, most important, by replacing the functional departments with a multidivisional structure. He expanded the research labs, modernized the offices, developed a new plant in Louisville, and worked to diversify the product lines. When doctors advised him of a heart condition, Young Otto decided to appoint Vincent L. Gregory, Jr., his successor and stepped down to become chairman in 1970.

The retirement of F. Otto Haas marked the end of an era. The baton passed to a group of able managers from outside the family, men who had come up through the ranks. For sixty years the Haas family had directly shaped the company's corporate culture and its strong commitment to innovation. Through subsequent decades F. Otto and especially John Haas, the younger son, would exercise a quiet influence in maintaining high standards of expectation and behavior. Meanwhile, founder Otto Haas was long remembered for his vivid talents as a business visionary, equally at home as a hard-driving taskmaster and a warm fatherly figure.

F. Otto Haas, shown here in an oil painting by James A. Fox, inherited a business that had grown from a modest partnership into a multinational chemical company with annual sales of more than $200 million and offices around the globe. His ten years at the helm made up a decade of change and continuity: he sustained the tradition of a corporate culture built on mutual respect and trust, as instilled by his father.

> As agriculture revived, farmers and vineyards would need modern insecticides and fungicides to control pests, kill mold, and improve the yield. Dithane Z-78 would spearhead Rohm and Haas's agricultural business in Europe, serving as a model for other products. . . . Rohm and Haas had to develop an efficient distribution system that would get the fungicide into the countryside.

Murphy sent blueprints for a Dithane grinder to the Feuchy plant manager, along with two engineers to help build the unit. Accustomed to the metric system, the French had difficulty reading the blueprints. The American engineers—one had been educated in France—calmed the waters and got the grinder up and running. Building the Dithane unit was the first of countless collaborations between European engineers and those from Bridesburg.[17] Murphy savored his victory—before moving on to the next challenge.

Americans in Paris

The next step was to set up a Paris subsidiary to oversee the entire European business. An experienced resident manager was needed to deal with banks, lawyers, Economic Cooperation Administration officials, French ministries, import licenses, export fees, and currency devaluations. There was another important reason for opening the Paris office: the customers. As Tetzlaff reported, "Based on my own experience in the European market to date, it is clear that we should be closer to the actual customers in the various countries."[18]

Up to this time London had been the center of Rohm and Haas's European operations. After George Kirsopp died in 1951, Tetzlaff initially orchestrated sales from Lennig's offices near Charing Cross in London. Lennig imported American-made products, and a trio of technical salesmen—Dr. Hope, M. H. J. Villeneuve, and J. L. Humphreys—promoted them to distributors scattered around Europe, but mostly in England. However, Murphy agreed with Tetzlaff and believed "that the outlet for our products on the continent will . . . be much more important than the U.K. market."[19]

As agriculture revived, farmers and vineyards would need modern insecticides and fungicides to control pests, kill mold, and improve the yield. Dithane Z-78 would spearhead Rohm and Haas's agricultural business in Europe, serving as a model for other products. Prochinor had to produce adequate quantities—and Rohm and Haas had to develop an efficient distribution system that would get the fungicide into the countryside.

Paris became the launch pad, with Tetzlaff moving from London to run the operation. In early 1952 he and Murphy signed the papers that established Société Minoc, the new French subsidiary, and divided French distribution among four companies. A promising young manager, Vince Gregory, was sent to France and, working out of the Scribe Hotel, helped set up the plant and locate office space. With the aid of Auby, Minoc secured quarters in an apartment building at 3, avenue du Président Wilson, near the Place du Trocadéro. Given the shortage of space in Paris, Gregory had to improvise, turning a kitchen and pantry into Minoc's first office.[20]

Murphy managed European growth from Philadelphia's Washington Square but was a frequent visitor at company sites around the world, including those in France, Spain, Italy, Switzerland, England, Holland, and Latin America. When transcontinental flights were still new, he thought nothing of spending twenty-seven hours on a plane, flying from Philadelphia to meetings in Latin America and then on to Europe. After the staff in London or Paris dropped him at his hotel, Murphy would unpack, shave, and shower; order

Published from 1943 to 1995, the Reporter *was an important vehicle for building trust with the customers, who were among the major readers. Articles often described their experiences with particular Rohm and Haas products. This issue from 1955 shows a tomato crop being harvested for an H. J. Heinz Company cannery. The crop had been treated with Dithane M-22 fungicide. Eventually the* Reporter *appeared in English, French, and Spanish editions.*

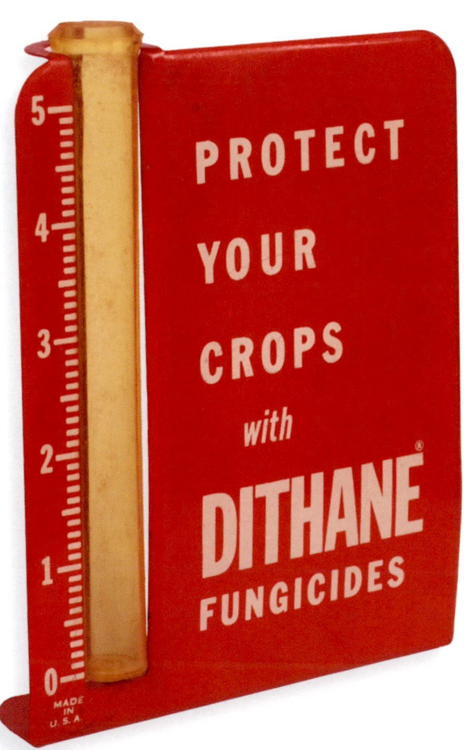

Rohm and Haas's subsidiary Société Minoc manufactured and marketed Dithane Z-78 in France, where young manager Vince Gregory, shown at right in 1952, helped the company navigate the French market. Salesmen gave customers small gifts that advertised Rohm and Haas products, including rulers, belt buckles, toy trains, and rain gauges like the Dithane example below. Opposite: On Pier 96 South in Philadelphia, circa 1957, stevedores wrestle drums of Dithane Z-78 into position on a Japanese freighter in preparation for the long journey to the ports of Kobe and Yokahama in Japan.

a bottle of scotch; and then buckle down to work, even on a Saturday or Sunday. He would operate nonstop for two or three days, going to bed at midnight and returning to the office bright-eyed at 7:30 a.m. His boundless energy astonished the office staff members, who were accustomed to the leisurely European pace.[21]

Murphy's work ethic and expectations sometimes conflicted with European customs. It took a while to reconcile the differences. For example, Elisabeth Timper, a Belgian secretary fluent in French and English, was astounded when her superiors asked her to take parcels to the post office and translate long technical documents. In the hierarchical French workplace, she told Otto Haas, highly valued executive secretaries had specific job duties, supervising the clerk typists and office boys.[22] Emulating Murphy, most of the international crew did whatever was needed to finish the job, and they expected the clerical staff to follow suit.

Minoc's managers soon learned both about the distinctive characteristics of the European environment and about how realities varied from country to country. As a licensee of Rohm and Haas's French patents, Minoc by law had to "manufacture as much material as the French market requires and . . . sell to all those who apply for the product." A vendor denied products by Minoc could appeal to the French government for a "compulsory license." In the United States no authorities told companies to whom they could or could not sell products. Another surprise lay in wait for Tetzlaff, when a Dutch customer advertised the fungicide under its own trade name, without mentioning Dithane, Minoc, or Rohm and Haas. Experiences like these challenged the Americans, who were accustomed to a different set of conventions.[23]

After its big sendoff Dithane Z-78 did not sell as expected. Although Dithane was effective on fruit, potatoes, tomatoes, and flowers, Mother Nature worked against it in wine country. When an unusually dry season kept the grapes free of disease, vineyards did not want to bother with an expensive fungicide. Tetzlaff blamed Minoc's sleepy sales agents. "The French distributors with whom we are currently working do not appear to have the usual selling characteristics to which we are accustomed in the United States," he wrote in 1953. "We feel sure that . . . Dithane can be established in the French market if the product is actually sold."[24]

Minoc's managers reevaluated the distribution system when Pechiney resurfaced, again asking for Dithane. The Pechiney Group had extensive research facilities at its disposal, including the Station Expérimentale de la Dragoire, a five-acre farm outside Lyon, and an auxiliary site at Villefranche. And in the French colonies—French North Africa, French West Africa, Madagascar, Vietnam, and the Antilles—it had superior technical service and marketing. Recognizing the Pechiney Group as players in the big league, Minoc agreed to sell them Dithane as the active ingredient for their Fongifruit and Fongicide Pechiney. To secure direct sales to farmers, Tetzlaff dropped Minoc's

DITHANE
Z-78
ROHM
HAAS
ON 234
YOKOHAMA
ASTORIA MARU

A SPIRIT OF COOPERATION

Rohm and Haas fosters a collegial environment at its research facilities, making the company a desirable home for the world's brightest specialty chemistry talent. The company's Springhouse campus, shown here, is located outside Philadelphia.

In the mid-1950s Rohm and Haas advertisements for Amberlite ion-exchange resins featured a series of oil paintings by artist Stanley Meltzoff. These dark, moody images showed a range of applications, from water purification to glycerine processing, using artistic license to suggest the Amberlites worked in mysterious ways. The close-up of the advertising art shown above—featuring a jar of Amberlite, a laboratory flask, and storage tanks—appeared in an advertisement in the July–August 1953 issue of the Reporter.

Give Us Ca or Mg

On June 24, 1948, editors from a wide variety of trade publications came to Philadelphia for lunch at the Downtown Club and a trip to the Rohm and Haas research labs in Bridesburg. They were eager to learn about the Amberlites, a family of ion-exchange resins expected to revolutionize scores of processing industries.

The press corps members received a warm welcome from attorney Edgar Kirsopp and were then taken to hear a talk by research chief Ralph Connor. Robert J. Myers, head of the applications lab, presented information about ion-exchange chemistry and the unusual properties of the Amberlites. The afternoon's highlight was a demonstration by ion-exchange specialist Robert Kunin. In rapid-fire experiments Kunin percolated brightly colored solutions through glass columns containing Amberlite beads and collected effluents of entirely different hues to illustrate the power and versatility of this separation technology. Sales engineer James C. Winters wrapped up, explaining how Amberlite absorbents could be applied to new fields like biochemistry, pharmaceuticals, and medicine.

In yet another illustration of Otto Haas's entrepreneurial interest in all varieties of specialty chemicals, Rohm and Haas first brought ion-exchange technology to the United States in 1939. In that year the Resinous Products and Chemical Company purchased the American patent rights from the two British inventors of these resins, Basil Albert Adams and Eric Leighton Holmes. The war put a damper on commercialization, but the company refocused on the innovation afterward. "From 1945 on, we put substantial research effort into the development of improved products," Otto Haas told the board of directors in July 1950. "In consequence, we have developed a range of resins, which, taken as a whole, is superior to that offered by any other company in the field. Our chief competitor is Dow Chemical Company."

Robert Kunin oversaw a growing research group in the Bridesburg lab that studied ion exchangers and their applications. Ultimately, Kunin won for himself a leading reputation among researchers in the field worldwide, thereby enabling further sales.

Ion-exchange resins turned into a solid, core business and one that Rohm and Haas repeatedly reinvented to suit the times. Soon after their introduction the Amberlites were mainly used to remove calcium and magnesium, the "hardness" ions, from the water needed by breweries, canneries, and laundries. Today they are used to purify liquids in academic scientific experiments, the nuclear industry, pharmaceuticals, biotechnology, and electronics. They are also used in food processing, where ion-exchange resins are applied in areas as varied as sugar refining and the removal of the bitter taste from orange juice.

TOCIL
東京有機化学工業株式会社
CHEMICALS
INDUSTRY
TOKYO ORGANIC CHEMICAL IND.,LTD.

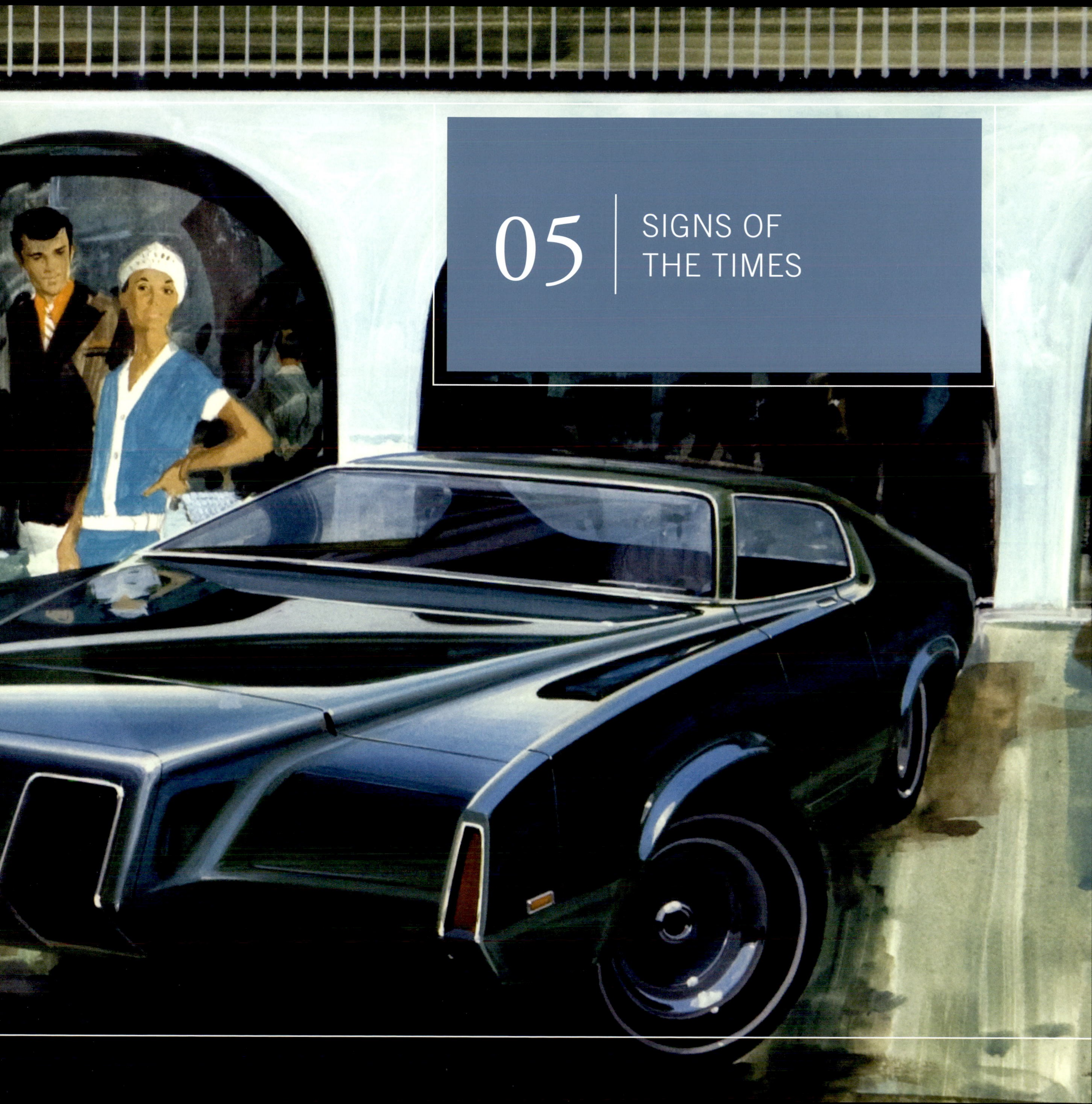

05 SIGNS OF THE TIMES

ROHM & HAAS
CHEMICALS FOR INDUSTRY
DEPARTURE
PLEXIGLAS
LEATHER VINYL
PAINT
IMPLEX

Darlings of Detroit

In January 1965 at Cobo Hall in Detroit everyone ogled the Explorer III, a Ford Mustang Fastback with an apple-red paint job—and several unusual new Plexiglas features. It was the annual meeting of the Society of Automotive Engineers, and the Rohm and Haas display was a showstopper. Explorer III was its latest "idea car," created to demonstrate the automotive design possibilities for Plexiglas plastic, Oroform elastic film, Rhoplex emulsions, and other synthetics.[1] By now a much larger company, Rohm and Haas was learning new ways to wholesale its products to large customers. Explorer III was the third concept car produced through a collaboration between Rohm and Haas and William M. Schmidt Associates, a design studio in Detroit. Schmidt had produced cars and trucks for Packard, Ford, and Chrysler before starting his own company in 1959. His most celebrated design

AFTER WORLD WAR II, ROHM AND HAAS SOUGHT TO FILL THE VOID IN PLEXIGLAS SALES LEFT BY THE DROP-OFF IN MILITARY USE OF THE MATERIAL. IN THE 1960S THE COMPANY CREATED A SERIES OF "IDEA CARS" CALLED THE EXPLORER. STOCK VEHICLES WERE MODIFIED TO SHOW HOW PLEXIGLAS AND OTHER PRODUCTS COULD BE USED IN AUTOMOTIVE DESIGN. THE EXPLORER I, SHOWN HERE, WAS A MAJOR ATTRACTION AT THE 1963 MEETING OF THE SOCIETY OF AUTOMOTIVE ENGINEERS IN DETROIT. ITS MANY PLEXIGLAS PARTS INCLUDED THE TAILLIGHT ASSEMBLY, THE INSTRUMENT CLUSTER, AND THE DECORATIVE MEDALLIONS. PREVIOUS SPREAD: THIS ARTIST'S RENDERING, CIRCA 1970, IS THE ONLY KNOWN SURVIVING RECORD OF WHAT MAY HAVE BEEN THE EXPLORER VI.

The 1960 Cadillac Eldorado, above, had taillights and door panels molded from Plexiglas and sported exaggerated tailfins to give the illusion of piloting a rocket. As advertised in the May–June 1955 issue of the Reporter, *right, Plexiglas molding powder was suited to the manufacture of freezer door panels, steering-wheel cups, range control knobs, taillight lenses, and refrigerator nameplates.*

ADMIRAL

Refrigerator Nameplate

Plexiglas ...the distinctive touch for fine products

Molded parts like those shown above combine functional value with gleaming beauty because they are made of PLEXIGLAS. This acrylic plastic has outstanding resistance to breakage, discoloration, weather and corrosion.

The combination of rich, brilliant appearance and rugged durability is the reason PLEXIGLAS acrylic plastic is chosen by manufacturers to give added sales appeal and serviceability to their products. You find parts molded of PLEXIGLAS, for example, on cars, home appliances, outdoor lighting fixtures, optical equipment and industrial pumps. Our brochure "Molding Powder Product Design" tells how and where to use PLEXIGLAS. We would like to send you a copy.

CHEMICALS FOR INDUSTRY

ROHM & HAAS COMPANY

Washington Square, Philadelphia 5, Pa.

would be the gadget-laden Batmobile in the *Batman* television series. For the last few years Schmidt had been in his element helping Robert L. Gardner, an executive in the Rohm and Haas plastics department, promote synthetic materials to automakers. For the Explorer I, displayed at the 1963 Society of Automotive Engineers meeting, Schmidt gave a facelift to a Chevrolet Corvair, adding a Plexiglas rear window and skylight along with headlights and taillights. Soon Bob Gardner had his satisfaction when all Detroit was emulating his designs. "In 1963, we introduced wall-to-wall taillights on the modified Corvair," Gardner told a reporter, and "two years later, the Buick Electra and Skylark came out with wall-to-wall taillights."[2]

Following World War II, Plexiglas rapidly became established as a design feature in American vehicles, initially as a weatherproof material for reflectors and taillights. As automakers became familiar with molded plastics, Rohm and Haas promoted Plexiglas for headlights, hood ornaments, nameplates, steering-wheel medallions, speedometer dials, instrument panels, and interior lighting. Thousands of rocket-shaped tailfins—the 1950s symbol of Detroit's fascination with the Space Age—were tipped with taillights in deep red Plexiglas. When jet motifs waned in the next decade, the physical properties of Plexiglas—moldability, transparency, and heat and shock resistance—ensured that it would still have a place in the automobile stylist's toolkit.[3] Designers, engineers, and executives expected to see creative uses for acrylic plastics and other synthetics—and Rohm and Haas responded with the Explorer series of concept cars.

Cold War, Hot Science

The name Explorer—taken from the U.S. satellite launched in 1958 in response to the Soviet Sputnik—resonated with Americans. The cold war was a patriotic age of big science: of universities, government agencies, and corporations together creating a golden age of R&D, coinciding with an era of unbridled economic growth. Between 1945 and 1960 the U.S. economy expanded very rapidly. For the first time ever, a majority of American households tasted a middle-class standard of living. The year 1960 marked a "great divide" in U.S. history, separating a period of relative scarcity from the new era of relative abundance.[4]

The United States emerged from World War II as the undisputed leader in the chemical industry, retaining this position until the late twentieth century, when globalization and changed economic realities began to shift the balance of power. From the 1950s to the 1990s, R&D generated breakthroughs in plastics, coatings, pharmaceuticals, fibers, and electronics, all

of which had enormous ramifications for consumers, the chemical industry, and Rohm and Haas as a company committed to research and innovation. By 1960, 11 percent of the employees of what had become a much larger organization worked in R&D, with the goal of finding and perfecting chemicals for many types of customers. "Our research organization is . . . looking for new and improved products which may be used in other industries," explained research director Ralph Connor. "We may hope to stumble on treasure, but the most reliable way to achieve worthwhile results is to decide what we want to achieve and then try to find a way to do it."[5]

Although not as large as DuPont, Union Carbide, or Dow Chemical, Rohm and Haas had distinct strengths. As the first mover in acrylics, it dominated this field, finding new Plexiglas markets and in due course developing a revolutionary acrylic-resin emulsion for quick-drying paints. In contrast, attempts to diversify into synthetic fibers, veterinary products, and pharmaceuticals would lead only to dead ends. Later, a fresh generation of managers seeking to revitalize the company's entrepreneurial spirit would find the road to renewal in electronic chemicals.

A Modern Material, New Applications

After World War II, Otto Haas consolidated the three companies that had resulted from his entrepreneurial zeal—Rohm and Haas, Charles Lennig, and Resinous Products—under the umbrella of Rohm and Haas Company, listed on the New York Stock Exchange in 1949. This move opened the way to streamline operations and to raise capital for expansion. The consolidated firm faced a new set of challenges: how to do business as a single company and how to use existing expertise in a dramatically changed world. One immediate challenge was to sell Plexiglas in a civilian market. During the war Rohm and Haas had run training programs at the army-navy air depots, teaching servicemen to repair and replace bullet-damaged windshields. When Plexiglas became available for civilian applications in 1944, thousands of repairmen who knew how to handle the material and understood its remarkable properties set up their own plastics shops. Pent-up consumer demand and the plethora of fabricators generated a postwar market for Plexiglas novelties.[6]

Plastics chief Don Frederick responded quickly, selling Plexiglas sheeting to these small businesses. One such fabricator, Vargish and Company of New York City, built its entire sales-

Rohm and Haas believed the development of colored Plexiglas would expand the demand for civilian use of the material. This polychrome window at the Oak Cliff Savings and Loan in Dallas, Texas, at left shows the creative use of tinted Plexiglas. Transparent gray panels provided glare reduction, while the colored panels added decorative detail reminiscent of a Mondrian painting. Below: Don Frederick's sales department developed booklets like Plexiglas for Store Improvement *that explained how Plexiglas signs, display cases, and lighting fixtures created a modern appearance that attracted customers.*

Diversification Decade

For over fifty years Otto Haas had built the company by expanding existing lines while relying on his German connections and his own research laboratories to find new products, all with one eye on the customer. Acquisitions were limited to the purchase of Charles Lennig in 1920 and the formation of Resinous Products in 1926. When merger candidates came across his desk—as did a Schering pharmaceuticals division in the late 1940s—Haas thought twice about bidding, motivated by a strong distaste for debt.

When F. Otto Haas and Ralph Connor took charge in 1960, they moved in a different direction and expanded the firm through a series of acquisitions. Young Otto lacked his father's formative engagements with Otto Röhm and his corresponding confidence in the future of acrylic chemistry, the core technology that generated big hits like Plexiglas and Acryloid oil additives. "I didn't think that the acrylic molecule had so much in it. I thought it was going to be quickly exhausted," he reflected in a 1984 interview. "I believed we had to find something different to take up when the acrylic molecule was all used up." The development of Rhoplex paint emulsions was well under way, but nearly two decades passed before the research paid off. Haas and Connor instead were persuaded by the merger trend that characterized American companies in the 1960s. Identifying with that trend, they took measures to turn their tightly focused specialty chemical company into a conglomerate.

The biggest investment by far was in synthetic fibers, but Rohm and Haas also ventured into health products and medical equipment. In 1963 the firm purchased Warren-Teed Pharmaceuticals, a prescription drug manufacturer in Columbus, Ohio. The next year it added veterinary products with Whitmoyer Laboratories in Myerstown, Pennsylvania, which specialized in poultry medicines, and the Affiliated Laboratories of East St. Louis, Missouri. In 1969 the company turned to medical diagnostic equipment by creating Consolidated Biomedical Laboratories from a group of smaller labs. Managers are not always adept futurists, and these inopportune choices steered the firm off course.

When the fiber market collapsed after the 1973–74 oil crisis, Vince Gregory was the CEO who had to reevaluate all these subsidiaries. Although Warren-Teed and Whitmoyer made a profit, they did not fit his new and more Spartan vision of appropriate territories for his specialty chemical company. Accordingly, they were sold in the late 1970s, followed by Consolidated Biomedical in 1982. "I could see that Rohm and Haas had been caught up in the 1960s acquisition boom that swept across the chemical industry," Gregory explained in 1995. "They made far too many acquisitions, thinking the net growth of the 1960s was going to continue. The acquisitions just didn't all work out." However, the lessons painfully learned in the 1960s and 1970s have served the company well in subsequent decades.

During the 1960s Rohm and Haas made a series of acquisitions in the fields of health products and medical equipment. Warren-Teed Pharmaceuticals was purchased by Rohm and Haas in 1963, but this diversification into pharmaceuticals was eventually abandoned by the company. Above, a laboratory worker at Warren-Teed records data, circa 1967.

Rohm and Haas had to improve the price, performance, and versatility of Plexiglas to succeed in the civilian market. Large chemical companies were now in the plastics business, vying for customers with rock-bottom prices. To maintain the lead in acrylics the board authorized expansion of the Bristol facilities and the construction of a high-tech plant near Houston . . . to produce lower-cost acrylate monomers.

promotion plan for its line of picture frames and bill holders around the plastic's heroic war record. Across the country leading department stores—J. L. Hudson Company; R. H. White's; Higbee's; Carson, Pirie and Scott; Kaufmann's; May Company; and Strawbridge and Clothier—put the merchandise in arresting window displays along with such Plexiglas aircraft parts as the big bubble canopy from the Republic Thunderbolt or the nose from the Boeing Flying Fortress. They promoted it in newspaper ads describing new uses for the famous war plastic.[7]

However, Plexiglas novelties and giftware—picture frames, bedside lamps, soap dishes, and umbrella handles—could not sustain the giant plants at Bristol and Knoxville, the latter purchased from the government in 1946. That same year Frederick told the stockholders, "Many of our small sheet fabricators have gone out of business, as they were not operating on a sound basis." Convinced of the virtues of his product and envisioning much larger markets—in aviation, autos, lighting, radio cabinets, packaging, signs and displays, vending machines, jukeboxes, business-machine housings, and interior decoration—the chief plastics officer updated his sales strategy.[8]

Rohm and Haas had to improve the price, performance, and versatility of Plexiglas to succeed in the civilian market. Large chemical companies were now in the plastics business, vying for customers with rock-bottom prices. To maintain the lead in acrylics the board authorized expansion of the Bristol facilities and the construction of a high-tech plant near Houston, in proximity to natural-gas supplies, to produce lower-cost acrylate monomers. Frederick stressed the need to think creatively about new applications for the acrylates and reported that "the Plastics Department has been reorganized to put greater emphasis on the development of new uses and better sales promotion." Suggestions included "use in synthetic fibers, water base paints, plasticizers and glass fiber reinforced plastics."[9]

Frederick also noted the "wide demand for colored sheets for civilian use." Right after the war Plexiglas was only available in clear, colorless sheets. As an interim measure the company introduced colored coatings that could be stenciled or painted onto clear sheets. The real innovation occurred when chemist Stanton Kelton, Jr., son of Otto Haas's lieutenant, discovered how to add pigments to the plastic. "If Plexiglas could be colored, then it could be used in a wide variety of applications," he remembered. "But there were no methods of putting pigments into the sheet. . . . There was no information on the stability of the colored sheets to outdoor weathering. . . . There was just nothing, really."[10]

The customer need for color spurred innovation, pushing the Bristol labs to work at a furious pace. "The Sales Department," Kelton continued, "sold the Plexiglas with the color in it and then said to us, 'develop the formula and make sure it's right.'" The labs were expected to "match the customer's sample," which came as a fabric swatch or paint chip. "We had . . . colored applications being thrown at us. We had to find dyes and pigments that would produce the proper color. We had to find ways to polymerize the sheets in the presence of these dyes and pigments."[11]

During the 1950s and 1960s Rohm and Haas published a special annual report for the employees. The cover of the 1955 edition, above, shows a worker packaging red Plexiglas molding powder for shipment to customers.

TESTED METHODS

Rohm and Haas helps its customers improve their products by testing efficacy and suggesting modifications. Here, research scientist Henry Eichman monitors a robotic floor scrubber designed to test how various floor polishes stand up to wear.

After World War II Otto Haas's younger son, John C. Haas (left), went to work for the company and continued his friendship with Otto Röhm, Jr. (right), who came to the United States to see the much-expanded American firm.

insisting that no two customers had the same needs. They catered to the new fast-food restaurants that were spreading across the country. Important clients included McDonald's and Mister S, a now-forgotten hamburger chain. For both, Sign Crafters made Plexiglas signs a part of the total image package: the large revolving S for Mister S and the golden arches for McDonald's.[23]

A Future in Fibers

The Plexiglas business, along with agricultural products in the United States and around the world, began to generate cash for new investments. "By 1960 we had so much money," board member Robert J. Whitesell remembered, "that we couldn't see places" to grow in the existing businesses. Otto Haas had long been intrigued by synthetic fibers and by pharmaceuticals, exploring these areas with guidance from Ralph Connor. When a staff member formerly with American Viscose suggested that acrylic emulsions might be spun into a new fiber, the entrepreneurial Haas "wanted to do it, . . . was all for it." In a fateful move he took the company beyond textile chemicals and into the far different business of making and selling actual textile fibers.[24]

As board chairman in the 1960s, in the years immediately following Otto Haas's retirement and death, Ralph Connor assumed responsibility for expansion through "acquisitions of companies whose products or production facilities might constructively complement our own activities." Aware that "Mr. Haas always wanted to have a fiber," he pursued this new field with vigor, making this area into the company's largest new investment. Rohm and Haas followed dozens of other American chemical companies that had moved into this promising area.[25] Before World War II, DuPont had introduced nylon, the world's best-known synthetic fiber, and had later developed a whole family of fibers, becoming the industry leader. Synthetic textiles were wrinkle and stain resistant, machine washable, and stretchable, qualities that had great appeal in the push-button, convenience-oriented postwar culture.

Fred Tetzlaff was assigned to fibers in 1962 after further developing his management skills in a series of North American positions. "He was an enormously ambitious fellow," recalled his boss Louis Klein, who oversaw acquisitions. "He was always moving, talking fast." A gambler, Tetzlaff bet that fibers would become a big win.[26]

Behind the fibers venture was the concept that Rohm and Haas expertise in acrylic technology would provide a pathway to innovation. The strategy consisted of establishing a presence with textile mills for commodities like nylon and then generating profits from specialty fibers made from acrylic emulsions. "You can pretty much tailor-make the copolymers," explained John C. Haas, Otto Haas's younger son, who worked in the company and served for many years as a family representative on its board. "Theoretically you could come up with a lot of special fibers that had unusual properties." The research lab initially believed that a hard fiber would create a unique product with outdoor durability, but eventually attention shifted to stretch fibers.[27]

Challenges confronted the fibers venture from the beginning. Two early acquisitions—Rhee Industries, a

Don Garaventi began his career at Rohm and Haas as a salesman in the Resinous Products Division and eventually became director of the North American polymers business. Garaventi, shown here dressed in painter's whites for a company photo shoot, quickly rose through the ranks by working in foreign operations.

South of the Border

Otto Haas set the stage for the company's expansion in Latin America when he courted the Argentine tanneries before World War I. By the time he retired in 1960, Rohm and Haas had two South American plants, one run by Filibra Produtos Químicos Limitada in Brazil and the other by Rohm y Haas Compañía in Argentina. Throughout the 1960s Donald Murphy forged ahead in the region by establishing subsidiaries in Mexico and Colombia. By then the Latin American region, like other overseas sites, had become an ideal proving ground for up-and-coming managers.

In September 1969 Donald C. Garaventi, a young salesman who wanted to get ahead, approached his boss, Henry J. "Hank" Schneider, about a raise. "Don, if you want to see your career go faster," Hank responded, "we recommend you consider working in foreign operations. There are some new opportunities in Latin America. Would that interest you?" A Lehigh University–trained chemical engineer, Garaventi loved to travel and was a whiz at explaining technical products to customers. The new job as marketing coordinator seemed ideal for his skills.

Elated, Don called his wife Lois from the home office, catching her while she was cleaning the oven. An adventurous woman who had put herself through Northeastern University and left her home in Massachusetts to work as a Rohm and Haas chemist, Lois exclaimed, "Go for it!"

In the Coral Gables office that managed the Latin American business, parts of Latin America were still seen as the wild, untamed frontier. Don's cohorts told him to watch out. They warned him that Colombia was unsafe because there had been a lot of kidnapping. As luck would have it, the Garaventis were sent to Bogotá in early 1971 and on to Buenos Aires in early 1972. No sooner had Lois hung up her drapes than Don was appointed general manager of Industrias Químicas de Apizaco (renamed Rohm and Haas Mexico in 1973) and the family was off to Mexico City for four years. Don was exposed to every type of business. He oversaw facilities that made emulsions, the rice herbicide Stam, and Plexiglas, and was involved in a joint venture to make ion-exchange resins with local partners.

"I learned a lot of important lessons in Latin America," Don recalled. "One of them was no matter how high you go in a company, you should maintain direct, personal contact with the people in the firm and with the customers. If the marketing people don't believe in your product, you're dead. The marketing folks have to help the customer understand how the product can help their company. The manufacturing and research people need to know that you understand their issues, and they want to understand what's going on in the rest of the company. The direct personal contact really facilitates the communication that is needed for a company to operate successfully."

What Garaventi learned in Latin America was duplicated by other young managers in other territories and other periods. In Europe there were Vince Gregory and Pierre Brondeau, while Raj Gupta cut his teeth first in Europe, then in Asia.

rubber thread plant in Warren, Rhode Island, and Sauquoit Silk, a nylon yarn manufacturer in Scranton, Pennsylvania—had small, outdated plants. At Rhee it proved difficult to sell rubber thread in a declining market while at the same time running trials on Rohm and Haas's experimental stretch polymer. "Work continues on the new elastic fiber developed by our laboratory," Klein told the board. "The long-term possibilities are good, but it will take some time before this is commercially significant." Sauquoit collaborated with Rhee to make the new stretch fiber, while struggling to produce nylon that could withstand price competition.[28]

Synthetic fibers seemed to stall in 1965, just as Tetzlaff was promoted to vice president and general manager of a new fibers division. The board optimistically approved nearly $20 million for a nylon plant in Fayetteville, North Carolina, in the heart of the southern textile industry. "We believe that the Fayetteville facility will be . . . a base for future fiber operations," Tetzlaff told the board members, and "the expansion of our sales force will bring us closer to the fibers market." A spun-acrylic product, XFE Experimental Elastomeric Fiber, was being tested in the Sauquoit pilot plant, while "a market development group" and "technically trained personnel" collaborated with textile mills to see how the fiber fared in blended fabrics. "We have high hopes that XFE will show advantages over products now on the market."[29]

Architectural Marvels

While Rohm and Haas was going deep into fibers, the plastics department continued on a roll as it explored additional Plexiglas markets. Movie theaters were already using Plexiglas sheeting in lobbies and marquees, an application that pointed to the possibilities inherent in architectural installations. The Plexiglas sales-service laboratory encouraged such applications by providing needed technical specifications.[30] But architecture and interior design were beyond the expertise of anyone at Rohm and Haas, whether at Bristol or Washington Square. To learn something about these potential markets, the plastics department confidently turned to the experts.

"We worked a lot with industrial designers," Don Frederick recalled, "with Walter Dorwin Teague and Raymond Loewy and Gilbert Rohde and Henry Dreyfuss, and all that crowd." Like many designers, the staff from Raymond Loewy Associates loved to experiment with modern materials, using Plexiglas in their redesign of the Coca-Cola soda fountain dispenser. They did a rebranding study for Shell service stations, ultimately replacing the rounded Plexiglas orange-yellow Shell symbol with a square Plexiglas sign in cherry red, white, and lemon yellow. They also used Plexiglas in interior layouts for retailers

Rohm and Haas's new home office was a showcase for Plexiglas, making use of some 73,000 square feet of the plastic. Among the building's many Plexiglas features were a Shirley Tattersfield–designed abstract mural, shown in detail opposite, and giant chandeliers that hung in the lobby, above. Each chandelier, designed by Gyorgy Kepes, contains 2,087 Plexiglas rods and weighs 1,290 pounds. Left: In April–May 1950 the Reporter *showed Frank Lloyd Wright's design for the V. C. Morris gift shop in San Francisco. Wright constructed a translucent ceiling from concave and convex Plexiglas domes of different sizes.*

ROHM AND HAAS COMPANY | TODAY

AN ELEGANT TIE TO THE PAST

Rohm and Haas's home office in Philadelphia made creative use of its signature product Plexiglas when the headquarters was built in the mid-1960s. Here, Plexiglas chandeliers adorn the home office's cafeteria, just as they do the lobby.

Through the 1960s Plexiglas continued to fascinate designers as a material that represented a sleek, modern future. In 1962 it was used in the Space Needle, the monorail, and other elements of the Seattle World's Fair, above.

like Famous-Barr, a St. Louis, Missouri, department store. "We even hired Walter Gropius, the founder of the Bauhaus," Frederick remembered, in the search for new Plexiglas applications. Gropius headed a research team, whose report described the material's advantages to architects: lightweight and bendable, with low thermal conductivity, Plexiglas was available in a variety of forms and colors, giving unlimited scope to the architect and designer. Gropius explained how it was ideal for curved walls, unusual lighting effects, and decorative treatments.[31]

This investment in design expertise brought prestige to Plexiglas and exposed Rohm and Haas to wider possibilities. Stylish storefronts, merchandise cases, skylights, geodesic domes, shower stalls, office partitions, streetlights, and indoor lamps all grew out of these interactions. By the 1960s Plexiglas molding powder was the preferred material for flat ceiling lights, and the Holophane Company was one of the largest customers. "Acrylic plastic is a designer's dream," explained Holophane sales manager Gene Rae. "You can do things with it that you can't do with any other material," as it is easy to mold, withstands heat from the lamps, and does not break when dropped. Architects and builders came to value Plexiglas for everything from bulletproof glazing to weather-resistant siding. The Structural Plastics Corporation put it succinctly: "Today's building industry and plastics belong together."[32]

Plexiglas became a favorite material for futuristic designs at world's fairs. Seattle's famous Space Needle and monorail were glazed with Plexiglas. At the 1964–65 New York World's Fair, Rohm and Haas plastics appeared in the street-lighting fixtures, public phone booths, and countless buildings and signs. And at Expo '67 in Montreal the massive geodesic dome at the U.S. Pavilion had a Plexiglas surface.[33]

The ultimate architectural showcase for Plexiglas was the new headquarters for Rohm and Haas itself. Only a few blocks from Washington Square, the building was located at Independence Mall, a historical district that was home to the Liberty Bell, Independence Hall, and the First and Second Banks of the United States. A collaborative design in 1965 by Pietro Belluschi, dean of the Massachusetts Institute of Technology School of Architecture and Planning, and George M. Ewing Company, Philadelphia architects and engineers, the structure made use of the newest materials, including dark-brown corrugated Plexiglas panels installed as sunscreens on the outside. The interior featured Rohm and Haas plastics in the form of doorknobs, telephones, screens, coatracks, and Lightolier ceiling fixtures. Several commissions—backlit murals by Shirley Tattersfield and Freda Koblick, a sculptural column by Arturo Cuetara, and lobby chandeliers by MIT design professor Gyorgy Kepes—demonstrated how Plexiglas could be interpreted by fine artists.[34]

While sales continued to grow, the plastics department operated in a maturing market where even entrepreneurial energy and marketing imagination were at best palliatives. Sales totaled nearly $50 million in 1964 and rose to more than $75 million in 1969, but profits fell from 9.5 to 7 percent. With the economy in

This investment in design expertise brought prestige to Plexiglas and exposed Rohm and Haas to wider possibilities. Stylish storefronts, merchandise cases, skylights, geodesic domes, shower stalls, office partitions, streetlights, and indoor lamps all grew out of these interactions. . . . The ultimate architectural showcase for Plexiglas was the new headquarters for Rohm and Haas itself.

recession, sales dropped to $63 million in 1970, and the profit was below 5 percent. The department coped by taking measures to reduce costs and by turning again to the Rohm and Haas tradition of customer service, directing additional resources to advertising, distributor training, and direct consumer marketing.[35] Meanwhile, much hope and effort focused on fibers, "the next new thing."

From Fashion to Fallout

In October 1969 Rohm and Haas sponsored a New York fashion extravaganza to tell the world about its novel stretch fiber, Anim/8. Three thousand people in the apparel, textile, and retailing trades attended the event in a specially built geodesic dome at the Tavern on the Green in Central Park. Guests saw a runway show of Anim/8 outfits, wandered through a maze of fabric sculpture, and watched a film that explained the fiber's properties.[36]

Taking cues from the success of the plastics division, the fibers division in its turn had spent months planning this send-off. Marketing manager Mike Storti worked behind the scenes with fashion designer Deanna Littell—a ready-to-wear specialist whose clothes were sold by high-end retailer Henri Bendel—and a New York City public-relations firm to generate interest. "The results of our meeting with Mrs. Diana Vreeland, editor of *Vogue*, and Mrs. Margaret Ingersoll, fabric editor, are nothing short of sensational!" publicist Delores Cree told Storti. Vreeland "insisted on having two garments—the black bikini and multicolor knit jumpsuit—flown to Hawaii today" to be photographed for *Vogue*.[37]

Anim/8 was the brand name for XFE Experimental Elastomeric Fiber, a spun acrylic fiber so revolutionary that the Federal Trade Commission created a new classification for it. Anticipating a large demand, Tetzlaff secured board approval to erect a $5 million Anim/8 plant adjacent to the nylon facility in Fayetteville to start making the new stretch fiber. "The extraordinary chemical resistance of Anim/8, its textile hand [niceness to the touch], and ease-of-care properties," he told the press, "qualify it as the first really practical elastomeric fiber."[38]

Continuing in the specialty mode that had been Otto Haas's recipe for success, the company was hoping to carve a high-profit niche with high-performance fibers. Vince Gregory, in his new role as CEO, approved plans for expansion. In 1971 Tetzlaff introduced two more hoped-for winners: Formelle color-spun hosiery yarn, which eliminated the "bag and sag" in panty hose, and X-Static nylon yarn, which alleviated the static in wall-to-wall carpets and double-knit polyester apparel. The marketing push continued as the fibers division collaborated with the Italian-born New York designer Giorgio di Sant'Angelo whose Patternskins bodysuits, knitted from Formelle, were shown in *Harper's Bazaar* and sold by Bonwit Teller.[39]

Alas, fashion brouhaha was no substitute for tangible advantages in price and performance. The new fibers soon flopped, and the whole focus on fibers turned into a fiasco. The marketplace reality was that by the early 1970s large producers like DuPont dominated the industry. They enjoyed enormous technical expertise and market clout, enabling them both to compete on

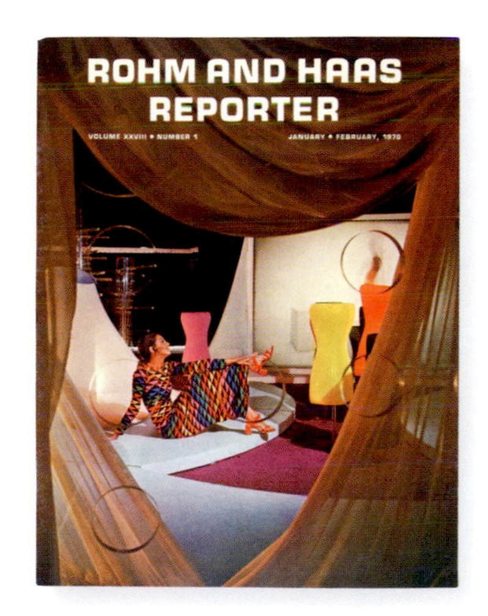

Designed by Deanna Littell, the mod dress featured on the cover of the January–February 1970 Reporter, *above, was sewn from fabrics made from Rohm and Haas's Anim/8 elastomeric fiber and was modeled at the Central Park fashion show of October 1969.*

The body-hugging quality of Formelle, a hosiery yarn developed by Rohm and Haas, was evident in the Patternskin bodysuits designed by Giorgio di Sant'Angelo in the early 1970s. Right, a display case at high-end retailer Bonwit Teller features mannequins clothed in the bodysuits as well as copies of Harper's Bazaar showing Patternskins.

price and to weather vicissitudes in the marketplace. Board members belatedly began to see fibers as "a big business—a business for giants." In a similar vein the Rohm and Haas nylon fiber program, by one outsider's estimate, appeared "to be a disaster." Tetzlaff clung to the specialty mantra, believing success would come in fibers as it had with Oropon bates, Tego adhesive, and Plexiglas canopies. "We believe we can be competitive with [smaller companies like] Enka," he explained, "even though we are not competitive with DuPont."[40]

The nylon venture continued to consume capital and made a profit only in 1973. Following that brief hurrah the world fiber market collapsed when feedstock prices skyrocketed and consumer demand plummeted in the wake of the mid-1970s oil crisis. Ready to retire, Tetzlaff passed the fibers baton to John Doyle, another manager trained in the international arena. It took a new generation of executives and an agonizing reappraisal by Vince Gregory for Rohm and Haas to come to grips with its flawed strategy and foundering investment.[41]

Don Felley, who ran the international division before becoming president and chief operating officer in 1978, summarized: "We came to the realization rather belatedly . . . that we did not have, in the fibers field, the critical mass to remain . . . through the thick and thin." Oil price hikes created "an entire new ball game. . . . It took the chemical industry some years to adjust. . . . It was a pretty bad time." The situation for Rohm and Haas was "exacerbated by the fact that we were in some businesses which we really shouldn't have been in, namely the pharmaceuticals and fibers business." In 1976 the company wrote off a $40 million loss on fibers in anticipation of selling the business in the following year.[42]

Heart and Soul

Rohm and Haas had tumbled headfirst into the new postwar order, wending its way through risk and opportunity by bold innovation. Otto Haas had fearlessly pointed to diversification; his immediate successors trod this path with mixed results. It was not easy for a specialty chemical company to find its way in maturing markets, where inventions like Plexiglas were fast becoming commodities and niche products like Anim/8 did not find a warm reception. Their contrasting fates provided necessary lessons to a company that was itself maturing.

The postwar Plexiglas market had required a different set of skills and a greater financial investment than military contracts for airplane canopies. In responding, the company built out from its scientific and manufacturing knowledge, learning lessons in how to develop fresh markets for known materials. For instance, using plastics in buildings was so new that most municipal codes did not provide for them. From the late 1940s on, the marketing and legal staff responded by lobbying for amendments to permit the use of Plexiglas in signs, auto parts, lighting fixtures, and architectural features. Collaborating with organizations like the National Safety Council and the American Society for Testing Materials, the company helped create standards that made acrylic plastics the preferred material for features like storm doors and shower stalls.

No sooner had the civilian market been established than the competition closed in. Maintaining a leadership position meant redoubling the research effort, strengthening technical service, building a strong distribution network, and reaching out to consumers. The Plexiglas sales-service laboratory extended the technical-service ethos to plastics, followed by the color laboratory, the plastics engineering laboratory, and extensive outreach programs for distributors and end users.[43] The expense strained the budget and eroded profits. Still, acrylic technology as manifest in Plexiglas was the heart and soul of Rohm and Haas: it was the backbone of the business at mid-century and remained an important part of the company's portfolio for many years thereafter.

Synthetic fibers were a different story. Here, the company tried to move into new and untested—although admittedly acrylic—technologies, while at the same time becoming a late entrant into markets with many large players. It was a losing proposition, and one that burned a hole in the firm's pockets. Tetzlaff tried to achieve a delicate balance between the production of nylon, a known commodity, and the development of esoteric and high-margin specialties. Even for a giant like DuPont with well-established markets the capital costs of fibers were high. For a new entrant and relatively small company like Rohm and Haas, the financial drain was devastating. "It's always hard for a small manufacturer to muscle in on the big guys," John Haas explained, "because they control the market through their advertising, through their volume, and so on."[44]

Rohm and Haas entered the fibers field too late; the qualities that were supposed to be unique to Anim/8 were soon imitated by others, "and that put the kibosh on it." The company learned the hard way that for a new entrant into a maturing field like fibers the usual selling points for specialties—the subtle differences that made one product preferred over another—were hard to establish in the minds of customers and mattered less than the price of what were being redefined as commodities.[45] With Oropon bate Otto Haas offered a product that the tanneries needed. With Plexiglas he had a unique product that was the fruit of Otto Röhm's special expertise and decades of development. However, with Anim/8 the company created a belated fiber that nobody wanted. Synthetic fibers dealt Rohm and Haas a blow from which the firm would take years to recover.

When the recovery did come, it was not from investment in a radical new market but rather from a slow steady build on the company's acrylic foundation, this time in the unlikely field of paint. The unique economic and scientific conditions that spawned the next revolution would test leader and lab teams alike but in areas they understood and in which they excelled.

In 1970 the plastics department launched its first merchandising program oriented toward the general public. Advertising appeared in consumer magazines and Authorized Plexiglas Distributors, above, were trained to deal with consumers' concerns. This strand of Plexiglas color chips, below, was used as a sales aid in much the same way as paint color chips are today.

K2

06 THE ACRYLIC ADVANTAGE

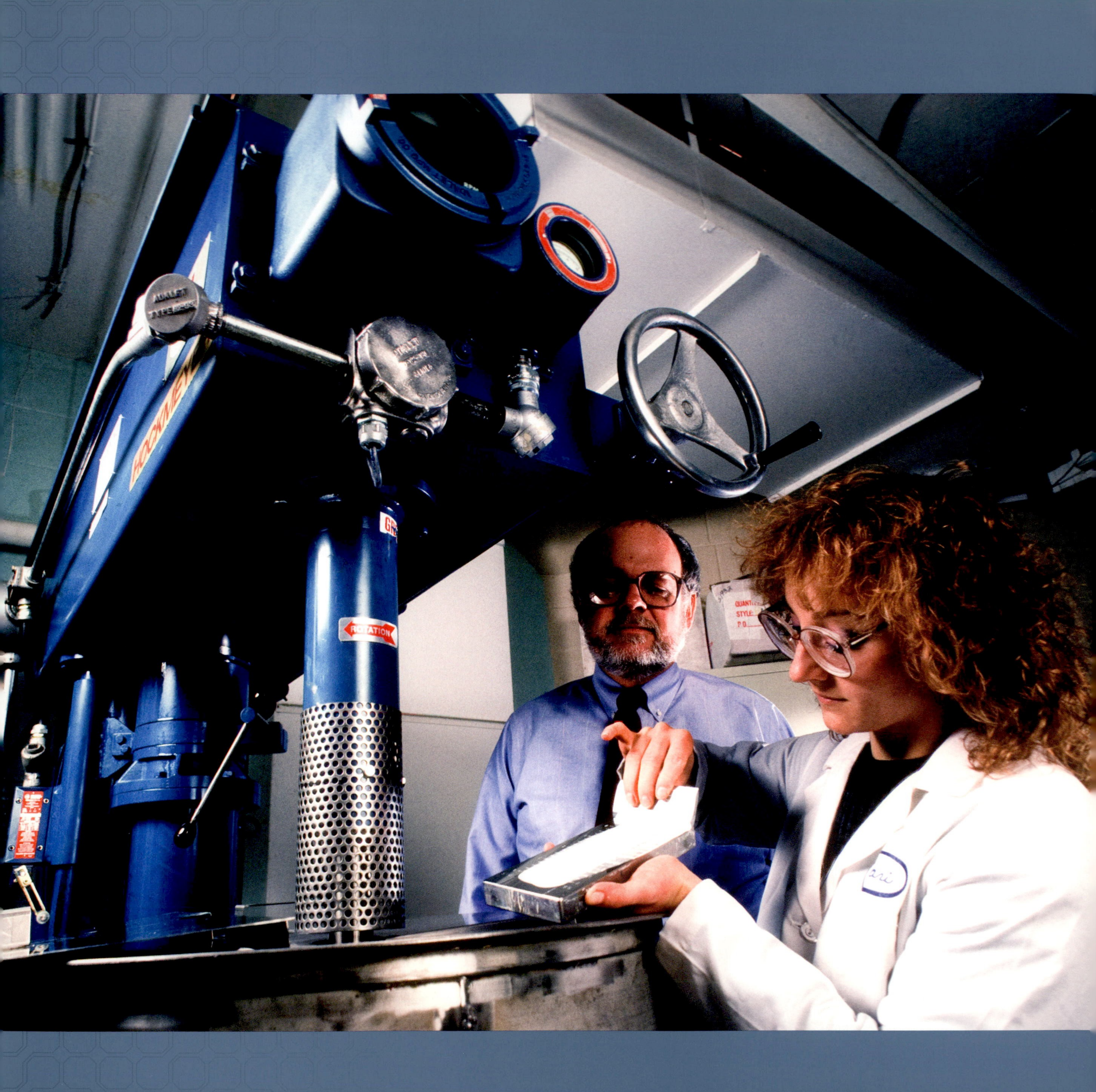
ROTATION

On Top of the World

"Our strong position in acrylic chemistry is one of the biggest drivers of our success. Everywhere around the world, everyone recognizes Rohm and Haas as the leader in acrylic chemistry."[1] By the start of the twenty-first century this proud boast was widely acknowledged. However, its basis laid not with Plexiglas—itself a commodity by the time it was divested in 1998—but rather in the very different territory of acrylic paints. The story of those emulsions involves complex relationships over a long period among issues as varied as investment in basic manufacturing, serendipity in research, and consumer aesthetics. The story reveals much about how Rohm and Haas has mastered the art of innovation through collaboration. In the years after World War II, Otto Haas, Don Frederick, and Stanton Kelton, Jr., succeeded in part by listening closely to their customers. That pattern was

ROHM AND HAAS'S ACRYLIC TECHNOLOGY IS USED BY MAJOR PAINT MANUFACTURERS THROUGHOUT THE WORLD. HERE, A COATINGS CUSTOMER PREPARES TEST SAMPLES OF DTM ACRYLIC GLOSS COATING, A WATER-BASED FINISH FOR METAL AND MASONRY SURFACES, CIRCA 1988. PREVIOUS SPREAD: THE HARSH WEATHER CONDITIONS OF THE EASTERN UNITED STATES ARE A PAINT SCIENTIST'S DREAM, WHICH MAKES THE PAINT QUALITY INSTITUTE FACILITY IN SPRING HOUSE, PENNSYLVANIA, AN IDEAL OUTDOOR LABORATORY FOR TESTING PAINT. PQI EXPOSES CUSTOMERS' ACRYLIC-PAINT FORMULATIONS TO THE ELEMENTS AT ITS "PAINT FARM," NOTING EACH SAMPLE'S RESILIENCE OVER TIME AND SHARING SELECT FINDINGS WITH CUSTOMERS.

Ralph Connor, shown above circa 1964, was Rohm and Haas's chairman of the board in the 1960s and helped steer the company through a period of expansion and diversification.

continued in the more complex world of the 1970s, 1980s, and 1990s. Ellington Beavers, a distinguished research leader, stated matters clearly: "Close integration of the company's strengths is absolutely vital to success in research and to the development of new products. You cannot develop new products in a vacuum, all alone in a research department, without the help of marketing and salespeople, and without contact with the customer."[2] Nowhere was this collaboration more important than in emulsions for acrylic paints. From initial steps to real commercial success would take twenty-five years, and a further quarter century would go by before Basil Vassiliou could make his claim.

Breakthrough Technology

After World War II, chemical companies scrambled for raw materials, putting pressure on production facilities. Otto Haas had long wanted to build a plant for "manufacturing some of our critical basic raw materials." When DuPont announced that it would no longer sell sodium cyanide—a key ingredient in the acrylic chain—Haas took action. In 1946 the Rohm and Haas board authorized construction of an $8 million plant in Deer Park, Texas, on the Houston Ship Channel. By mid-1948 Deer Park was synthesizing its own cyanide, then making intermediates and shipping them to Knoxville and Philadelphia. "We needed our own hydrogen cyanide in order to keep the bulk of our acrylic business, which was primarily based on MMA for use in Plexiglas sheets and molding powder," explained Lloyd W. Covert, vice president for production, "which all together represent a considerable portion of our business."[3] The Deer Park facility, Rohm and Haas's first substantial postwar investment in a new plant, foreshadowed the major expansion through acrylic technology that would develop outward from its Plexiglas beginnings.

Process improvements led to the slow buildup of the Houston site. Synthesizing the other important "acrylic" building blocks—the acrylate monomers, such as ethyl acrylate and butyl acrylate—for polymers was costly and complex. Bridesburg chemist Harry Neher sought new methods for their large-scale, economical production. Walter Reppe had earlier done similar experiments at I.G. Farben, but they were abandoned with the onset of World War II. Neher's lab had a breakthrough in 1949, developing a method of synthesis called the F process. "It is not yet possible to give an exact estimate of the degree of superiority of this new process," research director Ralph Connor excitedly told the board, but "it may be possible to sell acrylates at 70 percent of our current selling price. Market evaluations by our sales groups indicate that this lowering in acrylate prices would result in a substantial increase in their usage, and in addition make a significant difference in the margin of profit on the products we manufacture from acrylates."[4]

The discovery of the F process encouraged Otto Haas's natural optimism about his company's future. The company's acrylate polymer lines—leather and textile finishes—had limited market potential. Rohm and Haas's MMA products like Plexiglas directly used only small quantities of acrylate monomers, added as a softening agent. Rohm and Haas by now was master of

the acrylic universe, and with the F process it could make acrylates better than anyone else. "The great question," Haas wondered, "is can we sell the additional capacity," or could the company use the monomers itself?[5]

The challenge was finding new applications for the acrylates. As Haas later reflected, "If we look at our accomplishments in R&D during the period from 1946 to 1952, we must admit that most were not of the spectacular type." However, "we have some outstanding new products coming along—the most important being our acrylic paint emulsion." Faith in innovation was justified when markets began to respond to Rhoplex AC-33, an acrylic emulsion for latex paints, invented through a long, arduous process of trial and error. The invention "fits very well into our structure since we need additional acrylate usage or sales to support the F process plant. The additional investment required to make emulsions is relatively small."[6]

The first latex paints appeared shortly after World War II, evolving from government research on synthetic rubber. Based on styrene-butadiene or polyvinyl acetate water-based emulsion polymers, these paint systems had technical problems that prevented them from replacing traditional oil-based paints. They had poor moisture resistance and a tendency to yellow. Nevertheless, they attracted widespread attention because they were low odor, quick drying, water based, and easy to clean up.[7] Rohm and Haas made Aquaplex, an emulsified alkyd resin for quick-drying coatings. Aquaplex had to be discontinued before long, because of its instability. That failure left a sour taste in the mouth of Louis Klein, the vice president in charge of sales for Resinous Products, making him wary of paint experiments.[8]

For a long time rigid boundaries had existed between the work of scientists at Resinous Products and those at the legally quite separate company of Rohm and Haas. Otto Haas's 1948 merger of the two operations broke down the barriers. The board's subsequent authorization of construction for the F-process plant at Deer Park made the question of products and markets all the more urgent. The idea for acrylic paint emulsions came from two young Resinous Products scientists, Benjamin B. Kine and Gerald Brown. "Ben Kine was a very good chemist," recalled Charles McBurney, a research manager at the time. "He was making emulsions and supplying . . . leather coatings, textiles, and so on."[9]

Initial tests on house paints took place in Bridesburg, where chemists already worked on emulsions for textile and leather coatings. "We had some crackerjack paint technologists," Louis Klein remembered. Following the technical service model, researchers worked in collaboration with five paint manufacturers as they experimented on water-based (latex) paints. Scientist William C. Prentiss devised methods for measuring the characteristics of paint vehicles: adhesion, drying time, water sensitivity, soil retention, scrub resistance, sheen, freeze resistance, opacity, brushing qualities, and stability. He then reformulated paints from National Chemical, Glidden, Sherwin-Williams, Benjamin Moore, and American Water Proofing by adding some of the new emulsion "Rhoplexes." On the recommendation of Sherwin-

Sherwin-Williams promoted quick-drying acrylic paints to a new generation of do-it-yourselfers with booklets like this 1956 Home Decorator*, above. Young families living on a budget could spiff up their homes with elbow grease and a few cans of Super Kem-Tone and Kem-Glo latex paint made with Rhoplex emulsions. Left: In the tradition of sales plus service a salesman demonstrates paints made from Rhoplex emulsions to a customer in 1960.*

Williams and Montgomery Ward, samples of Kem-Tone and Kem-Glo were used as laboratory controls. Prentiss's research suggested that the Rhoplexes had great promise as paint vehicles.[10]

Prentiss outlined a research program for "the development of a superior latex for use in water-base paints." Research proceeded, with Kine and William R. Conn, who "had considerable experience with water-soluble or water-dispersible polymers," as leaders. Their experience, Prentiss explained, "can be utilized to develop superior modifiers for water-base paints." By spring Kine and Conn had landed on CL-135—made from ethyl acrylate, MMA, and acrylic acid—as a copolymer worthy of further study.[11]

The sales department could barely contain its enthusiasm, pressing the laboratories to move along. When hearing about the new emulsion, veteran salesman Felix L. LaMar exclaimed, "Fellas, I'm going to make you rich!" The lab supervisors put on the brakes. "Although Dr. McBurney and I have tried to make it clear that we cannot yet claim to have a new product," R. W. Auten told his boss, W. S. Johnson, the salespeople have "written and circulated a Sales Policy." CL-135 was to be market tested in two phases. "It is proposed that samples [from a pilot plant at Bristol] be submitted, on a confidential and research basis, to three carefully selected customers." Following these trials the emulsion would be subject to a "more comprehensive evaluation by nine more customers. . . . Thereafter we should know whether or not we have a unique emulsion suitable for use in the formulation of superior water-based paints for interior usage. Outdoor durability tests have been started."[12]

Progress stalled when researchers found that CL-135 yielded an emulsion with a low viscosity and a pink tint. The labs tinkered away, replacing the acrylic acid with MMA to produce good brushing qualities and making other adjustments to achieve a superior color. Samples of a new copolymer, ED 771, were shipped to a few paint manufacturers, while scientists worked with Bristol engineers to put the emulsion, given the trade name Rhoplex AC-33, into production.[13] A report late in 1952 indicated that "we have finally arrived at a composition which is being studied by several prospective customers. Three batches have been prepared without difficulty. The plant has been most cooperative in this work."[14]

In January 1953 Rohm and Haas announced Rhoplex AC-33 to the house-paint industry. "Over the past 25 years a few basic types of resins have succeeded in revolutionizing large segments of the protective coatings industry. . . . Rhoplex AC-33, an acrylic resin emulsion, seems destined to fill such a role in the field of water paints. Chemically, it is related to Plexiglas, the transparent acrylic plastic which has become a great new material of our time. Rhoplex AC-33 has many properties associated with Plexiglas—water-white color, chemical resistance, and resistance to aging. But it also gives the paint manufacturer the means of overcoming most of the limitations of earlier water paints."[15]

Doing It Yourself

Rhoplex AC-33 erupted like a volcano from the labs, after two years in the making. For the Resinous Products Division, saddled with aging products, the timing was perfect. In the area of older coatings and plywood adhesives Klein faced "ruthless competition" from "small operators" and others "with very little overhead." "Our optimism for next year stems from the fact that we

When Rhoplex AC-33 was in the development stage, several top executives, including research head Ralph Connor, had their houses painted with test paints made from the new emulsions. This 1956 photograph, opposite, shows a Rohm and Haas executive's home in Jenkintown, Pennsylvania, being covered with some of the new acrylic paints. Below: This 1966 brochure shows the famous advertising icon—the nine-year-old "Dutch Boy" modeled after the real-life Irish-American youngster Michael E. Brady in 1907—holding a can of a "new acrylic finish" formulated with Rhoplex.

CLEAN ENERGY SOLUTIONS

Rohm and Haas develops next-generation materials for the solar cell market then continues to improve upon them through rigorous testing. Here, research scientist Rebecca Hazebrouck runs electrochemical characterization for a copper plating bath.

ROHM AND HAAS COMPANY | **TODAY**

Pittsburgh Plate Glass Company, manufacturer of Pittsburgh Paints, appealed to the do-it-yourself crowd with promotional booklets, above bottom, explaining the easy application and beautiful hues made possible with acrylics. Acrylic paint was not just marketed for home projects: resorts like the Beachcomber Motel in Florida, above top, found that acrylic paints provided protection from the elements and that tourists liked the bright colors.

have a leading position in several new fields," he told the board in 1953. "We have some outstanding new products coming along—the most important being our acrylic paint emulsion. . . . We are putting tremendous effort into sales promotion of Rhoplex AC-33, our new acrylic emulsion for water paints."[16]

Despite AC-33's promise Rohm and Haas ran into difficulties in the marketplace. Some paint manufacturers looked askance at new ingredients or did not know how to use them. "We are still hoping," reported Auten, "that our customers will be able to formulate satisfactory Rhoplex-AC-33-based paints without casein," a milk protein used in water-thinned paints for thousands of years. "A few potential customers are being stubborn in this regard." Professional painters had learned their trade with oil-based paints and preferred to work with familiar mixes. When Ralph Connor had his house covered with test paints made with Rhoplex AC-33, the master painter handpicked a crew that could work with the new material. "There are some that won't be any good except in oil paint," he told Connor, "but these fellows will be open minded." Another obstacle was the composition of paints already in use. "By the time we were making Rhoplex AC-33, two-bit paint companies were making the alkyd resins" to use in their own paints, explained Klein. "It was easy to do." These firms did not want to invest in higher-priced ingredients from Rohm and Haas, even if they could produce a better paint. "We have made good progress," Klein reported to the board in 1955, "but the future is uncertain because even though technically superior, our product is substantially more expensive than competitive materials."[17]

One promising market for fast-drying paints emerged with the do-it-yourself movement that blossomed in the 1950s and 1960s. Budget-conscious young families saved money by doing their own plumbing, carpentry, gardening, and decorating. Inexpensive color effects could be achieved with a ladder, a brush, a roller, and a can of paint. User-friendly latex paint was perfect for the ambitious housewife or the "honey-do" husband who relaxed by fixing up the house. Acrylic paints, that is, improved latex paints, offered a superior product without going back to the hassle and expense of hiring, supervising, and paying a professional paint crew.[18]

Major paint manufacturers and national retailers were quick to identify this growth market. Forward-looking companies like Sherwin-Williams, Benjamin Moore, and Pittsburgh Plate Glass targeted the do-it-yourself crowd, and large retail chains like Sears, Roebuck and Company and Montgomery Ward introduced acrylic-latex paints in their stores and catalogs. Ward's 1957 spring-summer catalog described acrylic paints as the latest in science and technology, taking advantage of the Rohm and Haas reputation in plastics. Easy-to-use Wardflex paint, the catalog explained, was made from an "inert acrylic resin base (chemically related to Plexiglas)." With this "amazing new acrylic-base interior finish" thrifty consumers could "paint an average 12 x 14 ft. room for as little as $4.98."[19]

The larger paint companies, equipped with their own research laboratories, were the first to try Rhoplex AC-33. Montgomery Ward's chemists had already devoted five years to research on their own on latex paints when "the advent of Rhoplex provided the key to

One promising market for fast-drying paints emerged with the do-it-yourself movement that blossomed in the 1950s and 1960s. Budget-conscious young families saved money by doing their own plumbing, carpentry, gardening, and decorating. Inexpensive color effects could be achieved with a ladder, a brush, a roller, and a can of paint.

possible success."[20] Socony Paint Products Company, which made coatings for industry, used the acrylic emulsion to create a quick-drying interior paint for factories and offices. The British America Paint Company in Victoria, British Columbia, installed rooftop exposure racks to test exterior coatings made from Rhoplex. Benjamin Moore's impressive research facility in Newark, New Jersey, housed laboratories where scientists tried out Rhoplex along with other Rohm and Haas products.[21]

Even so, Rhoplex AC-33 fared best with custom-paint manufacturers like the Sinclair Paint Company of Los Angeles, which used acrylic emulsions in masonry coatings. A seasoned retailer, Frank Sinclair started producing paints for the military during World War II and later expanded his business by catering to the Southern California building boom. Since colonial times, stucco had been a popular exterior finish in the Southwest, where it was used on houses, apartments, dormitories, office buildings, and hotels. The material was a blessing and a curse. Buildings covered with this inexpensive, high-alkaline material needed frequent repainting. In 1954 Sinclair used Rhoplex AC-33 to create Stuc-O-Life, a masonry paint used on prominent buildings, including the Hotel Disneyland in Anaheim, the Maui Sheraton, and the Hong Kong Hilton. "We specialize in . . . customized paints to fit specific major architectural requirements," Sinclair explained. "In many cases we have on hand a formulation perfectly suited for the job. However, we frequently start from scratch and build the finish to meet the architect's specifications." Sinclair collaborated with architects, builders, and Rohm and Haas coatings experts to design special-purpose paints for waterfront housing developments on the Pacific Ocean and retirement communities in the hot, dry Arizona desert.[22]

Euphoria waned as Klein grappled with rising competition: "Sales are leveling off and it will take real hard work to continue to expand from this point." Large paint manufacturers learned to make their own emulsions, while rival chemical companies introduced competing products. One important latex paint customer deserted the acrylics for vinyl acetate copolymers, which offered lower quality and performance at a lower price. Price was an important factor for paint customers and paint users. Rohm and Haas sold $7.6 million of Rhoplex for house paint in 1960, but these more costly acrylic emulsions were still only a minor factor in the paint industry.[23]

The company ramped up R&D, focusing on small improvements that might give acrylic emulsions the competitive edge. Rhoplex AC-33 was terrific for interiors, but with the exception of stucco it did not perform well on exteriors. The laboratory introduced new generations of Rhoplexes: AC-34, AC-22, AC-388, AC-490, and AC-307. "A brilliant chemist named Eleanor Hankins discovered a monomer that could be added in very, very small amounts to the acrylic polymer," recalled Donald C. Garaventi, a retired executive who had managed the polymers business. "This was the beginning of AC-34, which had outstanding adhesion to wood and excellent durability but was still very economical."[24] Customers used the new emulsions to make paints that had different

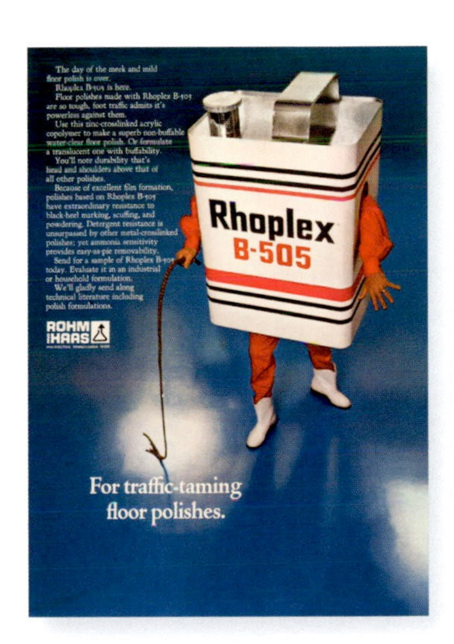

In 1969 Rohm and Haas produced a humorous advertisement targeting floor polish manufacturers, above, that touted the durability of acrylic emulsion Rhoplex B-505.

Rohm and Haas began its practice of exposure tests while still in its Washington Square headquarters. The Resinous Products and Chemical Company tested paints and varnishes on the roof of 222 West Washington Square, right. Opposite: Technical director David Stern (second from left) from the Martin Paint and Chemical Company, a Rohm and Haas customer, visits the paint farm in Newtown, Pennsylvania, in 1970. Martin used Rhoplex emulsions to make an extensive line of coatings for use inside and outside the home. When Stern visited the Newtown paint farm, his company had just introduced the "Lady Martin" line of scented paints.

sheens, were washable, repelled dirt, adhered to more surfaces, and resisted the elements.

The collaborative spirit that had been part of the company since the leather days shaped the innovative process. Garaventi, who had launched his Rohm and Haas career as a salesman in the Resinous Products Division, understood the role of technical service: "A customer tried a Rhoplex emulsion intended for paints with a different application. They used the emulsion to modify cement, to strengthen it for applications like patching compounds. Unfortunately, when they troweled it, it turned white because the emulsion was pulled to the surface. This problem was reported back to the chemists, who then made the emulsion in larger-sized particles so it wouldn't rise to the surface. This is just one example of how the creative process worked. The sales force understood what customers were trying to do and the problems they encountered. This information was reported back to research, and then the chemists analyzed the problem and came up with a solution."[25]

Exposure tests were conducted by hanging shingles on the roof at Washington Square and around the grounds at the Philadelphia plants. A new special-purpose exposure station—eventually relocated to Spring House—helped Rohm and Haas take acrylic emulsions from the living room to the front porch. The "paint farm" was a technical proving ground, where scientists collected data on how exterior paints endured the elements. It also was a meeting place for marketers and customers, who saw paint shingles weathering on the test fence—and knew that Rohm and Haas was paying attention to their needs.[26]

Down South and Over There

Acrylic coatings eventually became Rohm and Haas's best-selling line, surpassing Plexiglas as the star performer. "In acrylic emulsions, the first product [for paints] was introduced in 1953," recalled former CFO Fred W. Shaffer, "but it didn't have a significant impact on the company until the late 1960s."[27] By 1973 coatings were the firm's top product, accounting for one-fourth of nearly $400 million in sales and generating almost $13 million in profits. Rohm and Haas emulsions, led by AC-490, dominated the interior semigloss market, but, more important, by the decade's end acrylic paints had nearly supplanted solvent-based exterior coatings in the United States. "The North American coatings market," Shaffer told Vince Gregory in 1977, "has been and will continue to be our most important contributor to growth in sales and profitability." As one paint manufacturer put it, "acrylic is the magic word."[28]

Negley and Company, a San Antonio paint manufacturer, had spent decades trying to develop good paints for yellow pine, the favorite building material in the southern United States. The Texas climate brought out the lumber's worst characteristics. Humidity ranged from 15 to 90 percent, and temperatures could drop 40 degrees in an hour, causing the yellow pine to contract and expand. "We experimented with hundreds of oil-based paints," recalled general manager A. R. Rowe, Jr., "but the conventional systems we tested failed to yield the desired properties of flexibility and adhesion." In 1961 Negley began formulating paints with Rhoplex AC-34, creating the Crylicote line that "finally conquered yellow pine."[29]

The New Jersey–based NL Industries relied on Rohm and Haas technical service as it created Dutch Boy house paints. "There has always been a creative exchange of ideas and information," explained Dutch Boy's technical director Robert Bergfeld, "and this has been an important factor in our mutual success." Dutch Boy liked acrylic technology because it tolerated the harsh North American climate. "Under exposure to the

3 D
2-68 DA-6
2-68 DA-7
2-68 DA-8
2-68 DA-9
2-68 DA-10

Rockets at Redstone

In the 1960s Huntsville, Alabama, was a center of the space race and a fascinating blend of the old and the new. It had long been a cotton and textile center. During World War II it housed the Huntsville Arsenal, a munitions production site for the Chemical Warfare Service, and the Redstone Ordnance Plant, so named after the local red soil. Then in October 1948 the U.S. Army designated Huntsville as a hub for rocket research and development. The revitalized Redstone Arsenal, new home to the U.S. Army's Rocket and Guided Missile Agency and its Ordnance Missile Command, attracted high-tech contractors like Rohm and Haas.

Rohm and Haas established a Redstone Arsenal Research Division, sending nine scientists to Huntsville to conduct basic research on propellants for antitank weapons similar to the bazooka. "The company is undertaking this project as a public service," Otto Haas explained in the *Formula*. "We have no intention of entering into the manufacture of propellants. We were asked by the Ordnance Department . . . because we have within our organization a number of men whose wartime experience fits them to assume key positions in the technical organization." Among them was Ralph Connor, whose World War II experience included service with the National Defense Research Committee, the leadership group for vital wartime projects.

Connor described how the laboratory's goals fit with the company's collaborative model of innovation: "The Redstone Division is an organization with a variety of abilities and skills capable of handling a broad portion of the science of rocket propulsion. . . . Since the company's financial success is not dependent upon the operation of this division, we believe that we have an impartial position in the military propellant research picture. This has enabled us to enjoy excellent relationships with the Army, who contracts for our work, and with the other military and civilian agencies interested in propellant research."

In 1951 Colonel Carroll D. Hudson, Redstone's commanding officer, broke ground for the Josiah C. Gorgas Laboratory—a $1.5 million facility named after the Confederate chief of ordnance—and subsequently turned the structure over to Rohm and Haas. Over the next decade Redstone became the nerve center for army experimentation, as the focus shifted to propellants for long-range rockets and missiles. The Rohm and Haas staff grew to 240 people, and during the 1960s the division developed propellants for the Nike Zeus missile. In June 1970, however, the Redstone Division was shut down owing to lack of government funds, marking the end of the company's Space Age adventure.

In 1949, Rohm and Haas established the Redstone Arsenal Research Division in Huntsville, Alabama, where the U.S. Army had a research center that developed missiles and rockets. There, Project Nike began very late in World War II in response to the development of jet aircraft, which flew too fast and too high for then-current gun-based air defense systems to handle. Rohm and Haas developed propellants for all kinds of missiles, including the antiballistic missile Nike Zeus. Above, an early test model of the Nike Zeus missile of the type fired at the White Sands Missile Range in New Mexico on December 16, 1959, is prepared for takeoff.

varying temperatures and humidities characteristic of most areas of the United States, wood expands and contracts," explained Earl Beenfeldt, the marketing manager. "This places a lot of stress on a paint film and a lot of coatings fail," but acrylics endured.[30]

Domestic success was complemented by international growth. By the early 1970s the international division was one of the company's most successful departments, generating $278 million in sales and $28 million in profits, with Europe as the largest market. While agricultural products still dominated, divisional chief Don Felley and his staff worked hard to promote acrylic emulsions, sold under the trade name Primal. In countries with harsh climates Primal found ready customers—such as Tikkurila Ltd., Finland's largest paint maker. Like Americans, Finnish consumers had a five-day work week, spending Saturdays and Sundays on chores and hobbies around their homes and summer places. "A significant development . . . has been the increase in do-it-yourself painting in Finland," reported the firm's chief chemist, Henrick Furuhjelm. "Traditionally in our country, house painting has been done by professionals—or not done at all. Now things are changing." Tikkurila supplied more than 50 percent of Finland's paint needs and exported acrylic finishes to Sweden, Russia, and Turkey.[31]

Other European countries were slower to adapt. In Spain professional contractors did most residential painting, and the idea of the housewife with a paintbrush was just too radical. In the early 1970s changing values allowed paint makers like Industrial Bruguer of Barcelona to make inroads. "We in the paint industry have witnessed dramatic chemical developments in the span of a relatively few years," explained Bruguer's technical director, Cristobal Masso Feret. "These innovations have enabled manufacturers to develop truly revolutionary products that fully satisfy the needs of the consumer—and the conditions imposed by economic and environmental factors." Concerns over pollution, health issues associated with solvents, and rising labor costs paved the way for new approaches and technologies. In collaboration with Rohm and Haas, Bruguer created a line of acrylic house paints, aggressively advertised to the budding *hágalo usted mismo* (do-it-yourself) market in newspapers, magazines, and television.[32]

The $23.5 million riverboat Mississippi Queen*, above, made her maiden voyage in the summer of 1976. The new riverboat was finished with high-performance coatings supplied by Porter Paint Company and created from Rhoplex acrylic emulsions.*

The qualities that made acrylics sell in Finland and Spain did not resonate in places with little seawater or moderate temperatures. In Europe there was also stiff competition from chemical companies like BASF, which achieved economies by making both the polymers and the paints. "We didn't have much of an advantage over products that were cheaper," explained Vince Gregory. "We had to really do some good innovation to get products that could sell in the European market. We never did get the market share that we got over here." In Asia the Japan Acrylic Chemical Company, a joint venture with a local firm, also faced challenges. "Most of the Japanese output of emulsion paints is used for interior applications," the Far East manager lamented, "and hence the opportunities for expanding sales of our

ROHM AND HAAS COMPANY | **TODAY**

STAYING AHEAD OF TRENDS

Rohm and Haas manufactures materials that go into personal care products all around the world. Here, distinguished scientist Chuck Jones prepares a stress test for hair gel on a mannequin equipped with human hair and the ability to replicate natural head movements.

Sunscreens like Neutrogena UltraSheer Dry-Touch Sunblock, below, contain the Rohm and Haas product SunSpheres SPF Boosters, a technology based on opaque polymers. SunSpheres enhance the efficiency of UVB absorbers, enabling manufacturers to formulate higher SPF (sun protection factor) sunscreen and to deliver more aesthetically pleasing products. In the postwar years fine-art restorers experimented with Rohm and Haas products like Uformite 500 and Acryloid B-72 and discovered the advantages of synthetic resins. Today, conservators around the world use Rohm and Haas's Paraloid B-72 to preserve such priceless artworks as the Sistine Chapel, artifacts from King Tutankhamen's tombs, and the Terracotta Army of Qin Shi Huang, the first emperor of China, opposite.

Primal paint emulsions by exploiting their outstanding outdoor durability are quite limited."[33]

Opaque Polymers

While customer-focused product development led to incremental improvements, intensive research by creative scientists could produce groundbreaking technologies. When Donald C. Garaventi became director of the North American polymers business in 1978, his predecessor Allen M. Levantin told him about a promising endeavor at Spring House. "Don, they're working on something new out at the labs—opaque polymer. If you do anything to stop research on that, I'm going to shoot you!"[34]

An important additive in paint was titanium dioxide, which gave opacity to coatings but was very expensive. With the understanding that titanium dioxide made the paint white by blocking light, the paint industry searched for ways to imitate this effect. In 1974 Spring House launched a project to develop "more economic routes to opacity in coatings." Three years later there was a "technical breakthrough . . . based on hollow particle technology." One especially alert and persistent scientist, building on observations he initially made in 1965, brought water-filled styrene-acrylic particles into use. "Alex Kowalski," recalled Garaventi, "made emulsions, and saw that some had air trapped in the emulsion particles. He found that interesting, so he made more of them, and figured out how to entrap air in the emulsion particles. This gives opacity, without damaging the quality of the paint film."[35]

Opaque polymers appeared at a fortuitous moment, in the wake of the second oil crisis that dominated headlines in the late 1970s and early 1980s. The U.S. economy stagnated, with stagflation and a decrease in disposable income. Escalating prices for petrochemical feedstocks led paint companies to promote low-end products rather than the pricier acrylics. To counter these effects Rohm and Haas turned to its researchers. Two substantial advances—rheology modifiers and opaque polymers—offered hope. The opaque polymers, marketed as Ropaque OP-42, provided coatings manufacturers with a way to economize on raw materials without compromising performance. The shape and size of the particles—they were tiny uniform spheres—meant that paints could be formulated with less pigment, extender, and binder. Watching this business emerge and grow, Garaventi noted how the new polymer came about. "Ropaque resulted from beautiful chemistry, a serendipitous event, a brilliant chemist who knew what the market wanted, and a company commitment to find and develop meaningful new products. It showed that Rohm and Haas was willing to invest millions of dollars in research over a period of seven or eight years to perfect a product. In the end opaque polymers became a great commercial success."[36]

Over the next two decades Spring House researchers continued to address the changing needs of paint manufacturers. They spent the 1980s improving the performance of binders and additives and the 1990s responding to green initiatives for reducing volatile organic compound levels. New products like Rhoplex Multilobe, which fused small spheres into a lobed particle, gave exterior paints greater viscosity and better film properties, helping maintain the reputation for cutting-edge technology.[37] Marketers in coatings built on earlier work with Plexiglas and with acrylic emulsions for leather and textiles. North American homeowners became the focus of their efforts to broaden the public awareness of acrylic paints.

Motivating Consumers

In 1983 trade paint sales in the United States were split between contractors and do-it-yourselfers, with the latter buying nearly 60 percent of the product. Consumers mainly

Down Under

The Australian paint industry was born in the aftermath of World War I. In May 1918 British Australian Lead Manufacturers was established to produce white lead, a principal paint ingredient. By the 1930s the company was a dominant presence, selling a full line of paints and varnishes, including Duco nitrocellulose lacquer and Dulux enamel. These quick-drying coatings were used by automakers and other industries and by professional and amateur house painters, who appreciated their versatility and durability. In 1955 the firm was renamed Balm Paints, and in 1971 it became Dulux Australia.

As Rohm and Haas sought to expand emulsion sales around the world, Balm became an important customer for the Rhoplexes, which were sold under the trade name Primal outside the United States. The success that followed was due to the diligence of Dixon C. Van Winkle, head of Primal Chemicals, the Melbourne subsidiary that he and Don Murphy set up in 1959. Between then and 1975, this attentive manager turned a remote outpost focused on leather chemicals into a trendsetter in acrylic paints.

In the United States smaller paint companies were the major customers for the first acrylic paint emulsions, but the opposite was true "down under." The Melbourne office found its best customers among big players like Balm, the country's largest paint maker. Equipped with research labs, these companies could adapt the new technology to the local scene, working to overcome early technical challenges, such as excessive dirt retention, that eroded consumer confidence in latex paints. In 1964 Van Winkle reported that Balm was "the major factor in our emulsion sales figures, with their usage of Primal AC-34 in Lo-Gloss, an exterior house paint." Two years later he had "twelve steady customers, including the three largest paint companies in Australia."

By 1973 the Melbourne subsidiary, renamed Rohm and Haas Australia, happily reported sales of over $13 million, $4.5 million of which came from coatings. Dulux Australia was the most important Primal customer, accounting for 40 percent of emulsion sales. "The release of their new Weathershield Gloss and Flat paints," Van Winkle wrote in his annual report, has "assisted our improved performance." In second place was British Paints, which took cues from the industry leader. The small paint makers also began to take notice. "Whilst we will be relying on the big three customers, Dulux, British Paints, and Taubmans, to provide a solid base for the 1974 forecast, we have obtained substantial volume from many smaller customers which we have never enjoyed in the past. This should help further develop volume sales outside of the big three and lessen our dependence on them." The world was about to turn upside down.

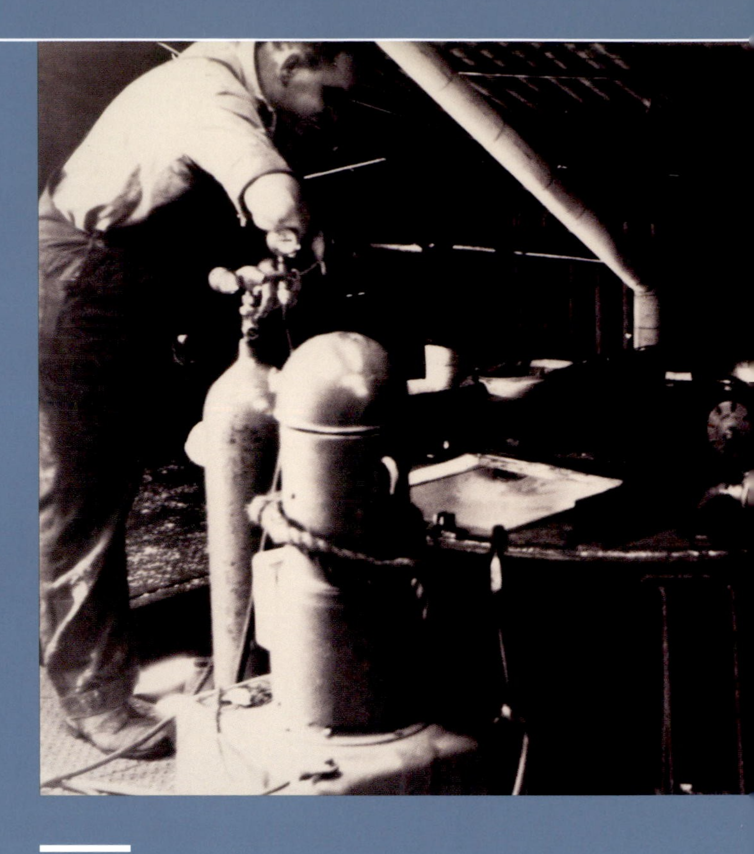

In 1966 Rohm and Haas subsidiary Primal Chemicals had a gross profit of nearly $429,000, more than 50 percent of which came from sales of paint emulsions. General manager Dixon C. Van Winkle happily reported on the "additional production capacity provided by the new 1,000-gallon kettle." A picture from a Primal scrapbook, above, shows a worker at the Geelong plant in Australia tending the new kettle in September 1966.

By 1989 Rohm and Haas materials were in 85 percent of the exterior acrylic paints sold in the United States. The two biggest customers were now Glidden and Sherwin-Williams, and large quantities of Rohm and Haas products also went to regional paint companies. . . . Ian Harris, a marketer known for his strategic thinking, learned about homeowners' needs by talking to these paint makers.

purchased from paint stores or Sears, but they sometimes shopped at discount stores and home-improvement centers, which expanded dramatically after the repeal of the New Deal fair-trade laws in 1975. On suburban strips big-box discounters like Kmart and Wal-Mart flourished by offering no-frills shopping to customers struggling to keep up with double-digit inflation. Cavernous warehouse stores like Home Depot targeted a new generation of do-it-yourselfers, baby boomers pursuing the American dream on a shrinking household budget. They wanted paints that were "more durable," but they were confused about how "latex" and "acrylic" helped a coating's performance.[38]

By 1989 Rohm and Haas materials were in 85 percent of the exterior acrylic paints sold in the United States. The two biggest customers were now Glidden and Sherwin-Williams, and large quantities of Rohm and Haas products also went to regional paint companies, such as Dunn Edwards in Los Angeles and MAB in Philadelphia. Ian Harris, a marketer known for his strategic thinking, learned about homeowners' needs by talking to these paint makers. Price wars squeezed profit margins, while such products as Union Carbide's vinyl acetate terpolymer threatened Rohm and Haas's lead in acrylics.[39]

A 1986 rebate program, which encouraged paint makers to put "100% acrylic" labels on their cans, had created some market awareness. Now the challenge was to get information in front of the people who bought the paint. Harris and advertiser Joe Sullivan approached Philadelphia-based public-relations consultant Al Paul Lefton Company. "One way to motivate painting contractors, retailers, and manufacturers to sell high-quality paint," explained Al Paul Lefton, Jr., "is to cause consumers to ask for it." Rohm and Haas had done this for an earlier product, through its Authorized Plexiglas Distributors, building a modest plastic-sheet business among hobbyists who loved to make their own furniture. The current need led to the Paint Quality Institute (PQI), a marketing unit aiming to educate homeowners on the acrylic advantage. Harris was soon dubbed "the father of PQI."[40]

Lefton initially expected to have a celebrity spokesperson run PQI, but some homegrown talent caught his eye. At a planning session he met Walter J. Gozdan, an affable scientist who managed the Spring House paint farm, now a six-acre site with twenty-five thousand outdoor test panels. Benefiting from two decades of experience, he knew the paint business inside out and could explain complex chemistry in a familiar, friendly way. The paint farm itself became part of PQI, which in turn launched a three-year outreach program to regional paint makers and consumers. Gozdan became a familiar figure on television talk shows and videotapes, in stores, and in *PQI Magazine*, a colorful new publication for the paint salesperson.[41]

In 1992 Rohm and Haas acquired the polymers business of Unocal Corporation, strengthening its position in vinyl acetate emulsions, which dominated the interior coatings business because of their acceptable performance for many interior applications and their lower cost. The Unocal emulsion acquisition gave Rohm and Haas a highly competitive position in

The Paint Quality Institute reaches out to paint sellers with products like PQI Magazine, *above, which provides tips on paints, interior decoration, and color trends.*

The PQI marketing team networks with trade associations and consumer groups to help explain the value of 100 percent acrylic paint. Above, PQI's Deborah Zimmer (left) speaks with visitors to PQI's Spring House, Pennsylvania, headquarters.

vinyl acetate emulsions to augment its strong acrylic position. The merger provided an expanded product line, more technology in vinyl acetate emulsions, knowledgeable personnel, and additional manufacturing facilities. Rohm and Haas was now a one-stop shop for both exterior and interior paint ingredients. The company had new territory to conquer—and a good reason to make PQI a permanent part of the marketing plan.[42]

Soon PQI found that even Gozdan's amiable sharing of knowledge had its limits. He would appear on a talk show to discuss the merits of acrylic paint but would inevitably be asked, "Walt, what color should we use?" In response Deborah Zimmer, a vivacious young marketer who had been at Rohm and Haas since 1980, was asked to take charge of interiors. Over the next decade she helped establish PQI as an advice center not only about durability, price, and performance but also about color schemes. By the late 1990s Rohm and Haas was strongly entrenched in the interior paint market, just as the boom in home-improvement television shows built consumers' color awareness. Zimmer tapped into these trends, adding her knowledge of chemistry, marketing, decorating trends, and aesthetics to the mix.[43]

Established to educate do-it-yourself consumers, PQI extended Rohm and Haas's sales-plus-service tradition to a new set of customers—painting contractors, paint retailers, architects, and journalists—using the latest media. Its educational films, CDs, training manuals, certification programs, magazines, and design contests were used to build an awareness of acrylic paint. "Our continued focus on training and education assists us in spreading the quality paint message," said John Stauffer, who spent thirty-eight years promoting architectural coatings for Rohm and Haas. By 2002 that message reached sixty million consumers per year through newspapers, mass-market magazines like *Good Housekeeping* and *Popular Mechanics*, and staff appearances on radio and television. The Web presented new opportunities. In 2006 more than three million visitors logged on to PQI's Web sites, making Rohm and Haas one of the world's most popular information sources about paint.[44] By the company's centennial year PQI was reaching out to the cyber generation through YouTube.

Zimmer and others marketers on the PQI team networked with trade associations like the Painting and Decorating Contractors of America and the Home Builders Association to share information. In turn the Rohm and Haas staff shared what they learned about the market with paint manufacturers to help them increase sales. "PQI is definitely unique in that its main purpose is to build our customers' businesses, which in return, helps build Rohm and Haas's business," said Pam Rogers, sales and marketing manager, in 2002. "Our new tracking methods show us exactly how our efforts impact the sales of a new product or affect a particular customer's sales. All of our research and feedback from customers prove that we truly do impact the bottom line."[45]

As the middle class grew around the world, PQI reached beyond its North American roots to new audiences in Europe, Asia, Australia, and Africa. Just as Basil Vassiliou brought an understanding of European culture to his ion-exchange customers, local experts on the global PQI team researched individual markets and oriented their messages and services to local audiences. "Many consumers are interested in color, style, and decorating ideas," said Linda Applestein, global director for PQI. "So, in addition to providing education about quality paint, we knew we would better serve the market by researching trends and providing inspiration on how to keep homes looking fresh and up to date. Then consumers have a resource to help them decide what paints to try, and retailers can make sure their shelves are stocked with what consumers want."[46]

Staying Focused

In many people's minds today the Rohm and Haas brand means acrylics. That leadership may be traced back more than a hundred years, to Otto Röhm's doctoral thesis on acrylic acid. His interest in acrylic chemistry led to Otto Haas's midcentury success with Plexiglas canopies and signs. These achievements laid the foundation for developments like Acryloid oil additives, Rhoplex emulsions, and Ropaque polymers.

When Otto Haas lugged Oropon samples around the tanneries, he established the Rohm and Haas reputation for customer care. By the moment of the "Old Man's" death in 1960 this service ethic was firmly embedded in the firm's culture. One example is the way that Don Garaventi spent his early career mastering the art of feedback and prospered as a result. "I viewed myself as an intermediary between the customer and the chemists. A success story comes when a customer has a problem, you solve it, and you end up selling the customer tank-loads of product. One of our jobs as technical representatives was to help our own technology people understand what the customer wanted. We made sure we did our part to have the scientists work on products that customers wanted, rather than creating something beautiful and elegant but not commercially meaningful."[47] Successful products came from the synergy between Rohm and Haas and its customers. Sometimes there was a eureka moment, but even an especially creative chemist like Alex Kowalski kept the changing needs of the marketplace in mind.

In 1998 and 2000 the Paint Quality Institute sponsored the Prettiest Painted Places contest, which celebrated communities that made creative use of exterior paints. PQI's goal was to increase the use of high-quality 100 percent acrylic paint. Cape May, New Jersey, above, was the contest's top town in the mid-Atlantic region in 2000.

But like all good strategies, even the "service first" ethic would be sorely tested as economic conditions, seismic shifts in raw materials, and technology breakthroughs forced Rohm and Haas to adapt quickly or run the risk of withering. Such wholesale change required conviction, a firm command of the issues, and the courage of a strong leader, in this case a business-savvy, former fighter pilot turned CEO named Vince Gregory, who in a few short years would both shake up and stabilize the company and the industry.

07 NEW TIMES, NEW VENTURES

Redefining Innovation

In 1970 Vince Gregory took on the job of chief executive with the grit of one who had known enemy combat and savored victory. His eighteen-year tenure as CEO coincided with a difficult era in Rohm and Haas's history. Acrylic technology remained the heart and soul of the company, but both Plexiglas and Rhoplex were beset with challenges. More seriously, the firm lost its way as the oil crisis of 1973–74 added new pressures to those that flowed from the "diversification fad of the 1960s."[1] Gregory handled the challenges with decisiveness, dignity, and fortitude. By 1970 Rohm and Haas was a larger, sprawling, international enterprise. Gregory and his successors, J. Lawrence "Larry" Wilson and Rajiv "Raj" L. Gupta, gave new meaning to the company's tradition of innovation. Innovation would no longer be exclusively associated with

ROHM AND HAAS CEO VINCE GREGORY, SHOWN HERE IN 1982, FOCUSED ON GETTING THE COMPANY ON TRACK AFTER A SERIES OF ILL-FATED INVESTMENTS WREAKED HAVOC WITH PROFITS. PREVIOUS SPREAD: ROHM AND HAAS'S RESEARCH AND MANUFACTURING FACILITIES IN MARLBOROUGH, MASSACHUSETTS, ARE LOCATED IN THE HIGH-TECH CORRIDOR THAT RADIATES OUTWARD FROM CAMBRIDGE INTO THE COUNTRYSIDE TO U.S. INTERSTATE 495. MARLBOROUGH MANUFACTURES MATERIALS USED IN SOLAR PANELS SIMILAR TO THE ONES THAT STAND ATOP ITS ADVANCED TECHNOLOGY CENTER BUILDING.

Shipley, the electronic materials manufacturer acquired by Rohm and Haas in 1992, opened a modern research and manufacturing facility, above, in Marlborough in 1982; that same year Rohm and Haas bought a 30 percent stake in the company. An example of early microchips is shown below.

product lines, novel technologies, and customer collaboration. Instead it slowly came to embrace ways of thinking, organizing, and getting things done. Deliberately managed organizational change was itself an innovation—and good for the company. The emphasis on customer service, integrity, and quality was still paramount, but the firm was given new structure and new direction. Gregory's willingness to adjust his management techniques as he guided Rohm and Haas through its most trying era was itself a new strategy and a model for the next generation.

A renewed emphasis on specialty chemicals led to promising fresh areas that would eventually become core competencies. The jewel in the crown would be the burgeoning field of electronic chemicals. In the 1970s electronics were still far from the nerve center of everyday life, but a new age was dawning. The military, government agencies, and large corporations used powerful computers to land men on the moon and keep track of the payroll, while consumers embraced transistor radios, hi-fis and stereos, microwave ovens, and tape recorders. Intel introduced the 1K chip in 1970, signaling the reality and future reign of Moore's law, and with it a remorseless shrinking of size, dwindling of cost, increasing of complexity, and growing versatility of application for the silicon transistor and the integrated circuit. Over the next few decades a wide array of electronic devices—calculators, video games, alarm systems, beepers, personal computers, portable TVs, CDs, DVDs, cell phones, and iPods—would become ubiquitous.

Specialty chemicals underpin the electronic age, providing essential materials to the microelectronics industries. These materials had to meet wholly new and ever more rigorous purity standards, while also being tailored to the highly specific requirements of each particular customer. Chemical compounds and processes that originated with startups like the Shipley Company of Newton, Massachusetts, enabled manufacturers like Intel and IBM to design faster chips and better computers. By 1980 a survey by the Charles H. Kline consulting firm valued this emerging electronic-chemicals market in the United States at $2 billion—not to mention global opportunities.[2] Gregory and his management team saw the potential of this novel territory for a company like theirs, with its long tradition of customer care and specialty chemicals, and responded.

Determined not to repeat the mistakes of the rush into fibers, in 1982 Gregory cautiously purchased a 30 percent interest in Shipley—one of the most respected names among printed-circuit-board and semiconductor manufacturers. Together the partners built on the Shipley footprint in electronics and expanded its already major presence in Asia. This was how innovation through collaboration happened at the new Rohm and Haas. However, before he could turn to this fresh territory, Gregory had much work to do in refocusing a firm that was ill prepared for a more competitive, globalizing world.

Stick to Your Knitting

Gregory understood the need for a healing process: the rebuilding of a specialty company that had lost direction. A Management Committee was charged with determining the future shape of Rohm and Haas. Discussion of human health products, veterinary products, and synthetic fibers dominated the early meetings as the team tried to reconcile the performance of these subsidiaries with market realities and expectations. Managers for each of the major business groups were asked to identify growth areas, regardless of whether the company had R&D efforts there. Caution ruled, however. "There will be no action," controller Fred Shaffer wrote on behalf of the Management Committee, "to pursue any of these opportunities until the fibers operations are sold."[3]

The firm's first chief executive with an M.B.A., Gregory brought professional management to the company. He made tough calls. "We had not only expanded into product areas that we should not have been in," he said, "but there was also a general expansion, with the growth rates of the 1950s and 1960s expected to continue. So we had too many people." His Efficiency Effort—the Double E Program launched in 1970, the year he became CEO—led to the first staff cutbacks in the company's history. Gregory also seized on a management philosophy based on the market growth–market share formula developed by Bruce Henderson at the Boston Consulting Group (BCG). An aggressive R&D initiative was launched to develop high-margin products in the traditional lines.[4]

The fibers dreams of the 1960s had turned into folly in the 1970s. Gregory ruefully described the debacle to *Forbes*. "I listened to the glowing talk from our fibers group that we could isolate ourselves from DuPont and Celanese and the other big producers of polyester and nylon because we had specialized grades. Boy, it sounded great! Well, it went right down the goddamn rat hole. By 1975, you couldn't give the stuff away!" In a company report Gregory also noted, "Our illusions were shattered by a double blow from the worldwide disaster in fibers and the severe recession—both of which hit Rohm and Haas very badly." As DuPont and Imperial Chemical Industries well knew, success in fibers came from investing hundreds of millions of dollars in research, chemical engineering, and marketing. Rohm and Haas as a smaller, new player could never assemble enough scientific horsepower behind Anim/8 or X-Static.[5] The lesson came at a high price. Gregory decided to cut the firm's losses, selling the nylon and polyester plants in 1977. Fibers had eaten into the company's cash and confidence, and reigniting the creative energy would take time.

As Rohm and Haas's comptroller, Fred Shaffer, left, had the job of evaluating the strategic objectives outlined by Vince Gregory's management team and determining whether those plans would make the company more profitable.

Further challenges came from the aftershock of the oil crisis. Gregory reoriented the company yet again, aiming to improve the return on net assets (RONA) in the traditional businesses. A major restructuring distanced senior managers from day-to-day operations, introduced the matrix reporting system, and established strategic planning as a corporate function.[6] Shaffer, who had worked at the firm since 1960, described the matrix

CHILLED WATER
SUPPLY
250 mL
PYREX
Fisherbrand
LABORATORY PROTECTIVE
EQUIPMENT

The financial reports said it all. In 1975 the company hit rock bottom, earning a 3.5 percent profit on $890 million in sales. By 1979 sales were $1.6 billion; earnings climbed to 6 percent. No longer weighed down by fibers, pharmaceuticals, and veterinary medicines, the firm was free to explore the future. . . . The big challenge was to identify those strengths and find ways to capitalize on them.

system as "breaking down the fiefdoms. . . . People couldn't become insulated and run their own show." In international markets the locus of decision making shifted from local subsidiaries to larger regional organizations, preparing for global growth.[7] Gregory and his fresh, young executives applied what they had learned at the Harvard Business School and from consultants like BCG to transform an entrepreneurial firm into a professionally managed company.

Trained in economics, Gregory appreciated the numbers and relied heavily on senior finance people like Shaffer and Larry Wilson—a fellow Harvard M.B.A. who had worked as treasurer at the home office and as regional director for Europe out of London, and was now a vice president and director—to help navigate through the storm. "During the 1977–1979 period we will concentrate our efforts on managing our core businesses as efficiently and profitably as possible," Shaffer wrote in the Management Committee's minutes. By the decade's end the emergency was over, thanks to divestitures and belt tightening. "The turnaround was quick," remembered Wilson. "Another oil crisis began in 1979. By that time we were fine. . . . We could raise prices, because the business was reasonably strong. . . . We got killed in 1975, but in 1979 we were okay."[8] The financial reports said it all. In 1975 the company hit rock bottom, earning a 3.5 percent profit on $890 million in sales. By 1979 sales were $1.6 billion; earnings climbed to 6 percent.[9] No longer weighed down by fibers, pharmaceuticals, and veterinary medicines, the firm was free to explore the future. "We will build for growth in the 1980s," Shaffer reported, "by developing new businesses and broadening our markets based on Rohm and Haas strengths."[10] The big challenge was to identify those strengths and find ways to capitalize on them.

William A. Kulik was among those on Vince Gregory's management team who thought long and hard about the company's future. The son of a poor Lithuanian immigrant, Bill grew up in Brooklyn, New York, joined the firm in 1960 as a sales trainee in resins, and worked in Argentina and Colombia before taking charge of the North American agricultural business. Kulik had a flair for the dramatic and a knack for being where the action was.[11] As director of the New Ventures Business Team, he evaluated business opportunities.

When Rohm and Haas acquired Shipley in 1992, it was a world leader in chemical electronics. Opposite, research scientist Emad Aqad works on organic synthesis of an extreme ultraviolet photoresist at the company's Marlborough facilities. Bill Kulik, shown below at his desk at Rohm and Haas's corporate headquarters in 1992, evaluated business opportunities as the company's director of the New Ventures Business Team.

ROHM AND HAAS COMPANY | **TODAY**

AN INDISPENSABLE PARTNER

To maximize the benefits of its specialty materials, Rohm and Haas works with customers to improve the performance of their products. Here, chemist Lingyun Wei evaluates the performance of Enlight metallization on a customer's solar cells.

Blazer

In April 1980 Rohm and Haas received word that the U.S. Environmental Protection Agency (EPA) was permitting the commercial sale of Blazer herbicide, welcome news for the company—and for soybean farmers. Blazer was a remarkable killer when it came to troublesome weeds like morning glories and ragweed, which robbed the valuable soybean crop of light, water, and nutrients.

Fifty years after it introduced Lethane, Rohm and Haas was still an important developer and manufacturer of fungicides, germicides, herbicides, insecticides, and miticides. Most of these agricultural chemicals were used on high-value specialty crops, such as apples, bananas, potatoes, rice, tomatoes, and wine grapes. Blazer was the first Rohm and Haas herbicide developed for a field crop—and soybeans were *the* most important field crop. In 1979 some 630,000 American farmers planted more than 70 million soya acres and harvested a record 2.3 billion bushels, a yield valued at more than $14 billion. Half the crop was shipped out of the country, making soybeans the number-two U.S. export after aircraft.

When Rohm and Haas scientists tested compound RH-6201 at the Newtown, Pennsylvania, farm in 1975, they knew they had a winner. "It knocked out many major weeds," explained Vic Unger, research director for agricultural products. "Generally in our business things are not so black and white." Patent protection, effective for seventeen years, started in 1976. In Latin America the governments in Argentina, Paraguay, Ecuador, and Bolivia quickly granted the company permission to sell Blazer to soy farmers for winter 1977–78, followed by Brazil, Colombia, and Uruguay. However, Blazer was developed in the U.S. regulatory context that emerged after the creation of the EPA in 1970. Before Rohm and Haas could market the herbicide at home, it had to carry out a long, complex program to confirm the compound's safety and efficacy. Company scientists did a ton of research—toxicology studies, environmental impact tests, and soil residue analyses—required for the application to register Blazer that went to the EPA in December 1978.

"I guess you've all heard of Blazer," Vince Gregory told the six hundred people who attended his April 1981 speech to the Chem Club, a group of Philadelphia-area employees who periodically met for dinner. "You're going to hear more about it in the years to come." Soybean farmers loved Blazer because it was powerful *and* cost-effective. Jeff Myers, who cultivated six hundred acres of soybeans on a family farm in Indiana, appreciated the fact that Blazer was a postemergent herbicide, which meant it could be applied after the weeds started to appear. No visible weeds meant there was no need to spray. This feature saved Myers ten dollars per acre on pesticides. Blazer was one of many examples of a specialty chemical company at work: in Gregory's words, the herbicide was meeting customer needs with research-based products and "a resounding success" for Rohm and Haas.

Research chemists Wayne Johnson (left) and Colin Swithenbank (right) were important players in the development of Blazer, an herbicide that was widely used by soybean farmers in the 1980s. In 1980 the Reporter *spotlighted Johnson and Swithenbank's contributions in a series of articles on the herbicide's history, which also discussed the roles of salesmen, marketers, and toxicologists.*

New opportunities were urgently needed. In the earlier rush to invest profits and diversify, Rohm and Haas had made a number of non-fiber acquisitions—Warren-Teed Pharmaceuticals, Whitmoyer Laboratories, Consolidated Biomedical Laboratories, and Micromedic Systems—which one by one proved troublesome. The problems landed in Gregory's lap at the same time as serious concerns of a very different sort. A cancer cluster at the company's Bridesburg plant caused deep concern, while broader public outcry over other companies' dioxin contamination at Seveso, Love Canal, and Times Beach contributed to the growing perception that chemical was simply a synonym for toxic. Clearly, a lot of things had to be done differently.[12]

The New Ventures Business Team provided a compass as it took on the "heavy responsibility of developing and implementing research programs which would result in new growth opportunities and new core businesses on which the company's long term future would depend."[13] At the same time the corporate business department would evaluate "new investment opportunities."[14] Improving performance, sharpening the focus on core businesses, and looking for step-out acquisitions in specialty chemicals formed a triple mantra. The Management Committee targeted specialties as the "principal vehicle for Rohm and Haas over the next decade."[15] Earlier generations—Otto Haas and Don Frederick, and F. Otto Haas and Ralph Connor—had pioneered this theme, but Gregory had to revitalize and reorient it for changing times. Larry Wilson summed it up perfectly: "The plan was to stick to our knitting."[16]

Finding acquisitions that fit the plan fell to Kulik. What would it mean to be a specialty chemical company going forward? How could managers know when a specialty was becoming a commodity? How much research was needed to maintain an edge? What were the danger signals? What would make for a good acquisition, and what should be avoided? By early 1980 Kulik's group had "identified fifty-seven sectors which together represent the known specialty chemicals industry," had studied the ten most promising in depth, and had targeted five areas playing to the firm's objectives: graphic-art chemicals, specialty plastic additives, pesticides, adhesives, and electronic chemicals.[17] The one that was to prove most fruitful was electronic chemicals.

At the Marlborough facility Shipley researcher Tom Sutter, above, operates a machine that washes copper-clad panels in preparation for their removal to the clean room where they will be coated with liquid photoresist, circa 1997.

As old-line chemical manufacturers remapped their strategies, many looked away from traditional commodities and, to reduce their dependence on oil, eyed high-tech areas like electronics. The aim was to get a leg up on R&D by acquiring small, dynamic firms. DuPont CEO Edward G. Jefferson spoke of making "toe-hold" acquisitions in electronics that would build on his company's existing businesses in photoresists and electrical connectors. A merger with Shipley—with its expertise in electronic chemicals, plating-on-plastics, and metal finishing—looked attractive. In November 1979, DuPont announced an agreement to acquire Shipley, but in May 1980 it cancelled the acquisition, citing antitrust concerns.[18]

Kulik and Wilson, who now had general oversight of new ventures and planning, reacted quickly. Shipley was an established name in the printed-circuit and microelectronics industries, and it had a strong

The husband-and-wife team of Lucia and Charles Shipley, right, was often compared to that of Marie and Pierre Curie, creative partners who responded to the opportunities of their times. The Shipleys founded their company in 1957 to develop new chemical technologies for the electronics industry and began partnering with Rohm and Haas in the early 1980s. Opposite: A customer dips circuit boards into an electroless copper solution sold by Shipley. The solution metallized each board with a thin layer of copper. The boards were then unracked, dried, dry-filmed, imaged, developed, reracked, and electroplated with a thicker layer of copper and tin-lead alloy supplied by LeaRonal, a company later acquired by Rohm and Haas.

presence in Europe and Asia. IBM could not build a computer and Intel could not make a chip without Shipley products: they were indispensible. There were also other, less-tangible assets that made Shipley attractive. Its nature as a family enterprise and its commitments to quality and customer service were good matches with the Rohm and Haas legacy of innovation through collaboration, a fitting return to basics as managers looked to the future.

Think Positive

Lucia Shipley was the most respected woman in the electronics industry. She ran the company founded by her husband Charles R. Shipley, Jr., in 1957. An inveterate tinkerer, Charlie had his first breakthrough—the Cuposit Catalyst 6F, a one-step liquid compound that helped copper adhere to the plastic and metal areas of a printed circuit board—in the basement of the Shipley home in Auburndale, Massachusetts. Together the Shipleys pioneered positive photoresists, a type of imaging chemistry used to make microchips. Charlie was the creative force, while Lucia made sure the payroll was met every week.[19]

Among their friends Lucia and Charlie Shipley were often compared to Pierre and Marie Curie, the husband-and-wife team who had worked side by side in the laboratory and shared a Nobel Prize for their work on radioactivity. Lucia was a hard-driving businesswoman who balanced Charlie's playful inventiveness, and their complementary talents made great chemistry. Before starting his own firm Charlie worked for Farrington Manufacturing, a Massachusetts company owned by Lucia's family. Farrington's Charga-plate system—a precursor to plastic credit cards—was used by hundreds of retailers and was followed by Scandex, an electronic system for reading and tabulating invoices at the point of sale. Farrington went deeper into electronics in 1954 by acquiring Electralab, a Cambridge assembly shop in the mysterious new printed-circuit-board trade. Charlie's job as Electralab's manager awakened him to electronics, and in 1957 he teamed up with coworker Cal Isaacson to set up the Shipley Company in the latter's garage in Wellesley.[20]

The Shipley Company designed equipment for circuit-board manufacturers, but Charlie soon learned a lesson. "After two to three months it became crystal clear that our entire concept" was off mark. Rather than "specialty equipment," he recalled, the customers "wanted . . . special chemical products" that were reliable and easy to use. "There weren't any that were really formulated for printed circuits. They were using products designed for something else, so there were a lot of problems." Much like the two Ottos decades before, the Shipleys turned to chemical solutions to alleviate production puzzles that caused bottlenecks on the shop floor. The aim was a chemical process whereby printed circuit boards could be manufactured simply, efficiently, and compactly.[21]

To cut costs, operations were moved to the Shipley home, where Charlie settled into midnight hours in his lab while Lucia managed the business from an office in the basement. "Charlie worked so intensely," she recalled, "it was exciting to see the progress."[22] The

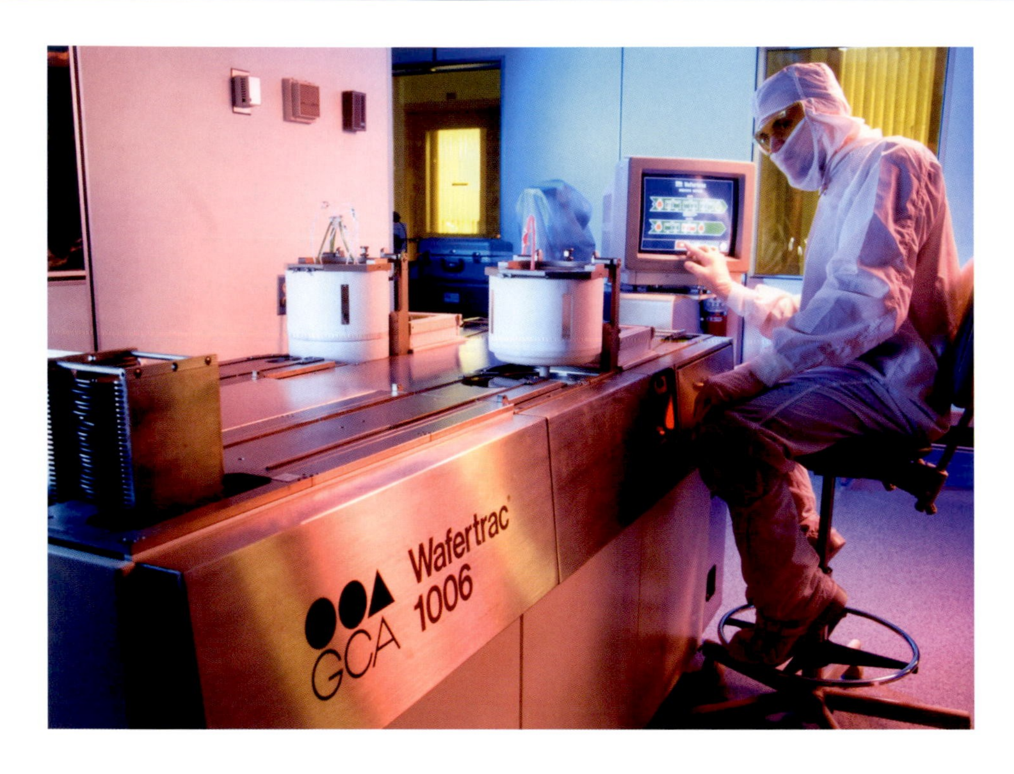

A technician, above, coats silicon wafers with photoresist on a GCA Wafertrac machine in Marlborough, circa early 1990s. The coating is done before the wafers are taken to the stepper, which transfers the mask pattern into the photoresist. The exposed wafers are then brought back to the Wafertrac to be developed. Silicon wafers, below, act as the substrate for microelectronic devices.

eureka moment came one Sunday morning in April 1958, when experiments pointed to Catalyst 6F for printed circuit boards. “The critical decision was whether to work on the breakthrough or to continue to sell,” Charlie recalled. “My wife decided that we should work on the breakthrough.” Field tests took place in nearby Waltham at Raytheon, a leader in defense electronics.[23]

When the Shipleys took Catalyst 6F to market, they found ready customers among small businesses like LaPointe Industries, a Connecticut manufacturer of television antennas and accessories, and technology leaders like Remington Rand, General Electric, and IBM. “For a number of years Remington Rand’s purchases amounted to anywhere from one third to one half of the company’s business,” Charlie remembered. He personally attended to sales, packing his Rambler station wagon with samples for the drive to IBM in upstate New York. If this sounds like Otto Haas carrying Oropon to the tanneries, the comparison is apt. Charlie relished these visits, during which he would demonstrate the product, train the customers, help them troubleshoot, and learn what they needed next.[24]

In one instance IBM declined technical assistance only to have one of its chemists, a man named Fisher, make a distress call a few weeks later. “I told them that I would be out first thing in the morning,” Charlie recalled, “and I drove all night. . . . Here is this poor chemist, alone in all of this sea of IBM people who were obviously mechanically inclined and couldn’t possibly understand why chemistry wasn’t in black and white.” When the problem got fixed, he “thought Fisher was going to kiss [him]!” Rescue missions like this taught Charlie about customers, products, and goodwill.[25]

Catalyst 6F revolutionized the infant printed-circuit-board industry, making it possible to manufacture reliable double-sided boards and adopt volume production techniques. A second innovation, CP-70—an electroless copper solution that plated over metal and plastic surfaces catalyzed with Catalyst 6F—eliminated labor-intensive steps in production and enabled circuit-board manufacturers to achieve further efficiencies. These two products and their successors became the basic chemicals used to make printed circuit boards, feeding the growing demand for high-quality electronics.[26]

The Shipleys began to expand their business into other electronic niches. Charlie was intrigued with the capabilities of the positive photosensitive resists used in offset printing, which he believed could be modified for the printed-circuit-board trade. David Arnold, an early Shipley employee, played a role in bringing the requisite technology to Shipley, networking with the German dyestuffs manufacturer Kalle A.G., a division of Hoechst (formerly part of I.G. Farben), while in Europe. Back in the United States, Dave and Charlie struck a deal with Azoplate Corporation, Kalle’s sister company in New Jersey, to provide raw materials for Shipley’s experiments. In a few years Shipley developed the first positive working photoresist for electronics, both for circuit boards and microchips. In 1963 Shipley entered an agreement with Azoplate and Hoechst, obtaining the rights to their materials for use in the electronics industry. The next year Shipley introduced AZ-1350, the only photosensitive resist capable of fabricating the high-resolution masks, or templates, used for transferring images to silicon chips. In the years ahead the semiconductor industry would use AZ-1350 to make ever-more-powerful chips, and Shipley steadily improved the line to keep up with the demand for better imaging materials.[27]

Shipley's plating systems had broader applications in the metalworking and plastics industries, since they deposited uniform metal coatings that adhered successfully to the complex contours of different objects, such as the inside and outside of a ballpoint-pen tube.[28] The first use of metal on plastics was in radio and television knobs, caps for salt and pepper shakers, and mixer housings. The technology soon found a larger market in the auto industry as lightweight grills, hubcaps, door handles, mirrors, headlights, and taillights. It was also used for high-tech applications in chemical engineering, wire-and-cable manufacturing, and the aerospace industry.[29]

By the late 1970s Shipley, now headquartered in Newton, Massachusetts, had manufacturing facilities and sales agents around the world. A California office served the needs of Silicon Valley, while sales offices and distributors in England, France, Italy, West Germany, Holland, Sweden, and Switzerland attended to the large European market. Dave Arnold had built up the European business, anticipating that rising living standards would "make it too expensive to employ all those people shuffling papers" and lead instead to wide use of mainframe computers. In 1978 a manufacturing plant opened in Niigata, Japan, to make products sold by Shipley, Japan (later called Shipley Far East Ltd.) out of Tokyo. From a tiny startup with $13,000 in sales, Shipley grew to a $64 million business by 1980.[30] However, all was not well.

Although Charlie continued to tinker, the Shipley Company "wasn't developing any new products," his son Richard later explained, and "was losing market share right and left. The semiconductor industry hadn't taken off. We were selling liquid photoresists to the printed-circuit-board business, but we were being crushed by dry film, and DuPont had a good dry-film business. If these problems weren't big enough, we were also losing our shirts on the deposition side of the business, where MacDermid was our big competition. Our revenues were skyrocketing because the market was going quickly, but we were getting our clocks cleaned." To cope, Lucia came out of semi-retirement to orchestrate the firm's future, joined by son Richard C. "Dick" Shipley, a Boston University M.B.A. who had worked at the family business on and off since his youth.[31]

Some of the minority shareholders wanted liquidity. The merger discussions with DuPont had collapsed. New York investment bankers were enlisted to find another buyer. Several chemical manufacturers quickly expressed interest, but the family wondered whether a merger would drain Shipley of its independence. They were much like Otto Haas, who cancelled the Shell deal after World War II for fear of

Rohm and Haas's research and development director, Robert E. Naylor, shown above in 1990, came to Rohm and Haas after a long career at the DuPont Company and helped orchestrate the Shipley acquisition.

CONTROLLED ENVIRONMENT

The precise nature of Rohm and Haas's electronic materials require pristine conditions only available in the industry's most advanced clean room environments. Here, a worker at the company's Marlborough facility tests photoresist on silicon wafers.

ROHM AND HAAS COMPANY | **TODAY**

In Gregory's second decade as CEO, Rohm and Haas performed better than almost any other chemical company in America. . . . *Forbes* ranked Rohm and Haas first among diversified chemical manufacturers worldwide in return on equity and earnings per share, as big producers like DuPont, BASF, Hercules, and Union Carbide retreated from commodities, slashed payrolls, and struggled to get into specialties.

INTERNATIONAL

Shipley Catalyst

THE SYMBOL OF INNOVATION

Major investment in new R&D building

Executive Seminars Focus on Technology Trends

Charlie and Lucia Shipley valued good communications and over the years issued a series of monthly publications for employees, including the Shipley Newsletter, Quality at Shipley, *and the* Shipley Catalyst, *above.*

losing control of his company. Charlie said, "Lucia, we've tried selling the business. It didn't work, and we're not going to sell. Dick's back in the business. Let's sell a minority interest."[32]

Rohm and Haas became the preferred partner, in part because the acquisition team—Kulik, Wilson, and Robert E. Naylor, the new R&D director—respected the family's desire to retain control. "Bill Kulik had been making an appeal to us to buy the company," remembered Dave Arnold, "and explained that they had chosen to get into specialty electronic materials, out of fifty-seven possible fields. They made a strong appeal to us. . . . Many times the uncertainties of the future with our supply from Kalle were almost a deal breaker. Kulik always maintained a very positive and optimistic view that things would work out."[33]

Quality Matters

As Rohm and Haas weathered the severe recession of the early 1980s, Vince Gregory proudly announced the Shipley acquisition. "The fabrication of microelectronic devices and printed circuit boards is an outstanding example of how specialty chemicals can contribute to technological advancements in the most rapidly growing industries." The company paid $38 million for 30 percent of Shipley, hedging its bets on a bright future for electronic chemicals. Gregory sold Consolidated Biomedical Laboratories and discontinued an oil-and-gas exploration project with Shell. The "diversification splurge" had ended.[34]

In Gregory's second decade as CEO, Rohm and Haas performed better than almost any other chemical company in America. In 1983 the stock price hit a record high after rising 150 percent in twelve months. Wall Street analysts were exuberant about the new management team. *Forbes* ranked Rohm and Haas first among diversified chemical manufacturers worldwide in return on equity and earnings per share, as big producers like DuPont, BASF, Hercules, and Union Carbide retreated from commodities, slashed payrolls, and struggled to get into specialties. "The plain fact is," said DuPont's chief economist, "chemicals aren't an exciting growth industry anymore." Nothing was guaranteed. Even niches like electronic chemicals, which promised to grow and grow, were not immune to the market's volatility.[35]

The new management techniques—strategic planning and the focus on returns—did increase productivity, but operational discipline came at a price. People worried that RONA was an autocratic, top-down system that, disturbingly, pulled the company away from its traditional pillar of strength, listening to the customers. The board concurred that the "emphasis on efficiency and centralized control . . . produced good results but was not conducive to risk taking, individual initiative, autonomy, and entrepreneurship." Rohm and Haas needed to "refine some basic corporate values and begin to use them." While the core mission remained—"we supply high quality products and good service to the customer"—there was confusion "as to who is responsible for knowing the customer's longer-term needs so that we can anticipate them."[36] The corporate office began to focus on people, customers, and innovation, while production managers explored ways

Larry Wilson, who became chairman and CEO of Rohm and Haas in 1988, was instrumental in moving the company into the electronic materials industry. Wilson, shown here in 1990, had joined Rohm and Haas in 1965 as an operations research analyst. He retired in 1999 after leading an industry-wide effort to promote corporate responsibility in chemical manufacturing.

Next Gen

In June 1989 *Chemical Week*, the industry's leading journal, sat up and took notice of Rohm and Haas. The company was "spending money, . . . more money than it ever had in a capital program—$1 billion over three years." The investment in new plants stood in contrast to the merger strategy of the 1960s and the belt tightening of the 1970s. A new generation of managers had just come on board, and they had distinctive ideas about how to balance change and continuity.

J. Lawrence Wilson took the reins from Vincent L. Gregory, Jr., to become chairman and CEO in June 1988. With John P. Mulroney as president and chief operating officer, Wilson was poised to build on Gregory's achievements, make the firm more market oriented, and strengthen its reputation in specialty chemicals.

Wilson arrived at Rohm and Haas when he took an entry-level job as an operations research analyst in 1965. He came with a B.S. in mechanical engineering from Vanderbilt University, experience with the navy Seabees in Bermuda, an M.B.A. in finance from Harvard Business School, and an appreciation for Rohm and Haas that was developed by reading the annual reports and sneaking into a stockholders' meeting. As Wilson climbed the corporate ladder, he spent time managing such Rohm and Haas subsidiaries as Warren-Teed Pharmaceuticals, Consolidated Biomedical Laboratories, as well as the European region in the wake of the oil crisis. From this experience—and from watching Gregory turn the company around to refocus on specialty chemicals—Wilson learned to be cautious in the best sense of the word. "I don't see any need for us to try to branch out," he told *Chemical Week*. "You have to stay where you have technical competence."

Wilson kept the focus on specialties, while also enhancing Rohm and Haas's leadership role in the chemical industry. Decades earlier the industry enjoyed wide esteem and could afford indifference to the occasional public expression of concern. However, Rachel Carson's 1962 best-seller, *Silent Spring*, cast cancerous suspicions on the chemical industry's record. Only slowly did managers awaken to the need for better public understanding. The Chemical Manufacturers Association's Responsible Care program, begun in 1988, was one central if belated response designed to improve the industry's health, safety, and environmental performance. Wilson played an active role in mustering support for Responsible Care. He received the Society of Chemical Industry's prestigious Gold Medal for these accomplishments, following his 1999 retirement. "We have a proven ability to develop new products and technologies that solve problems," Wilson told his fellow executives. "We have learned that responsible behavior is what will make a difference for future generations."

to improve the quality of the product: total quality leadership, or TQL, was in.

Gregory read the writing on the wall. "With competition coming, especially from the Japanese, we realized that we had to get all of our people involved." The management guru W. Edwards Deming, who had advised Japanese industry in its march to reliability and quality, was helping transform the ailing Ford Motor Company into Detroit's most profitable enterprise. Ford's total-quality-management program inspired other U.S. manufacturers to consider Deming's work. The drive was on to have statistics and teamwork lead to better American products and services. Quality became the watchword at firms like General Motors, Western Electric, Polaroid—and Rohm and Haas. As Gregory was to put it, "We have always been close to customers at Rohm and Haas, but our value has been extended to include the Deming philosophy and Total Quality Leadership."[37]

Enthusiasm for Deming percolated up from the Knoxville plant, where in early 1983 managers and engineers heard the master speak. Within a year everyone at Knoxville had been trained in the Deming way. Soon other plant managers embraced the approach. When Roy Djuvik, operations director for the North American region, and Sam Talucci, vice president and regional manager, got wind of these developments, they acknowledged that quality was "not just a manufacturing problem" but also "a total corporate issue." Djuvik explained: "We have to ask ourselves, 'How do we know that customers are getting what they want?' Statistics is one way to get answers."[38]

The electronics industry lived or died by quality, and TQL fit the needs of Rohm and Haas subsidiaries like Electro Materials Corporation of America (EMCA), which made chemicals for microelectronics. (To carve out a sizable electronics niche, the company had also acquired EMCA and three other firms—Furane, Isochem, and Plaskon—in the early 1980s). As an independent company, Shipley ran its own show, building on Lucia's commitment to meeting her customers' exacting specifications. "Quality is the essence of our business," she wrote, "and our reputation for quality is one of our most important assets." In 1984 Shipley launched a formal quality program called the Continuous Improvement Process (CIP), with the advice of Deming disciple Bill Conway. CIP eventually morphed into Total Cycle Time, focusing on the speed—in development, production, delivery, and service—that is so critical to the semiconductor business.[39]

Vince Gregory brought in management guru W. Edwards Deming, left, to work with Rohm and Haas senior managers to improve quality in the 1980s. Opposite: In Rohm and Haas's Cheonan, Korea, manufacturing facility, an employee takes a sample of a photoactive material made for the microelectronics industry to test it for quality, circa 2008.

Vince Gregory brought Deming in to advise senior managers and oversee seminars on "the philosophy of quality, with its emphasis on precision, performance, and attention to customers." Some 80 percent of employees, including hourly workers, received training. Like other prudent executives, Gregory did not blindly adopt Deming's ideas but added his favorite metrics to the mix. RONA still had a place, given its role in turning the firm around. At a stockholders' meeting John P. Mulroney, who became vice president and chief operating officer in 1986, explained the "improbable" marriage of RONA and TQL. RONA kept managers

John P. Mulroney, who worked at Larry Wilson's side as president and chief operating officer during the 1980s and 1990s, played to his strengths as a diplomat and cultural mediator, above.

focused on the money, while TQL combined statistical analysis, teamwork, and constant improvement to achieve "long-term customer satisfaction."[40] Here was Otto Haas's customer-centric philosophy updated for the professionally managed electronic age.

Shipley's Toehold

The Far East was the most exciting part of Shipley's international business. Back in the 1960s Dave Arnold had focused on the business in Europe, shying away from Japan owing to language barriers and laws limiting foreign investment. For Japanese distribution he relied on Ichiyo Kotani at Muromachi Kagaku Kogyo in Tokyo. Steadily increasing sales stirred Arnold's curiosity, and on a visit to Tokyo he found the entrepreneurial Kotani teaching small factories to make printed circuits, which launched this industry in Japan. A door opened in 1974 when a new law permitted 100 percent business ownership by foreigners. Seizing the opportunity, Shipley established a wholly owned subsidiary with Kotani as director, becoming one of the few U.S. companies with a toehold in Japan. "At the end of the first year the Japanese subsidiary was making money hand over fist and producing a profit. This very powerful and profitable Japanese operation became the forefront of Shipley's strategy in the Far East," Dick Shipley explained. "Kotani built a very strong operation. He worked with people whom he could trust, and who did a superb job for Shipley, including our distributor network in Korea, Singapore, Hong Kong, and Taiwan."[41]

By the mid-1980s the U.S. printed-circuit-board industry began to worry about foreign competition. In 1984, a banner year for U.S. board sales, Asian imports increased by almost 40 percent. Most U.S. electronics companies now assembled devices in Asia, and forecasters predicted that Japan would use more circuit boards than the United States and Europe in the second half of the decade. The electronics industry was going global.[42]

With its toehold in Europe and Asia, Shipley took advantage of these changes. By 1988, 60 percent of revenue came from outside the United States. The company had three sales offices in Southeast Asia, together with a technical service center in Singapore and a plant in Sasagami, Japan. George Reed, who had managed the offices in Singapore and Hong Kong, described the new global order. "For many years almost any goods or services produced in the United States were in demand everywhere," he explained in 1992. "Today, it's a much different world. . . . More creative approaches are needed." In Japan products were obsolete as soon as they hit the market, and Shipley needed to have fresh products ready to replace the old. "One of the keys to global success is the state of mind that continually looks beyond what is needed today."[43]

Growth through Reinvention

Vince Gregory had turned a troubled family company into a disciplined managerial firm by the time he passed the baton to Larry Wilson in 1988.[44] It appeared as if the company was set for fair sailing. However, higher oil prices during the Gulf War led the American and the wider global economy into the 1990–92 recession. These years were tough for all concerned, including Rohm and Haas and Shipley. The job of coping fell on

the shoulders of Wilson and his team. The bottom line still looked strong, but profits camouflaged a hard truth. "The biggest problem was that the company wasn't growing," Wilson remembered. "We had to figure out how we could grow."[45]

The focus on profits and quality had led to products and processes being put on the back burner. For instance, polymers and resins—one of the core businesses—made the same quantity of product in 1976 and 1986. No new construction had taken place for this business since the mid-1970s, and backlogs interfered with manufacturing efficiency. Plants worked overtime to keep up, sometimes buying acrylate monomers because Deer Park could not make enough. The company had plenty of cash, but in several areas old facilities and mature portfolios were turning what had been growing businesses into potential liabilities. Wilson's first response was to build new plants, ten of which were completed in 1990.[46]

Under the banner of continuous change Wilson unfolded his own growth strategy. The plan was to build a "defensible portfolio" by divesting unprofitable businesses, reinventing a select number of older ones, and developing electronic chemicals and other new specialties. Wilson acted decisively when the opportunity came up to acquire full control of Shipley, which was in dire financial straits as the electronics industry buckled under the recession. Substantial resources were needed to expand the semiconductor side of the business, and discussions about where to find the money had strained family relationships. Bob Naylor was sent to New England to help Lucia Shipley understand how a sale could benefit both sides.[47] In 1992 Rohm and Haas took full ownership of the garage startup that Charlie and Lucia had built into a Route 128 superstar.

Looking both to critical mass and to new specialties, in 1998–99, at the end of his watch, Wilson made his boldest stroke of all, doubling the size of the company by brokering a merger with Morton International. Morton was two-thirds a chemical company and one-third a salt company that owned the famous Morton Salt Umbrella Girl logo.

Acquiring Morton gave Rohm and Haas a "cash cow" in salt and several new specialty ventures. It also meant that the company jumped from a $5 billion mid-cap to an $8 billion large-cap: the world's leading specialty chemical company.

Vince Gregory's "stick to our knitting" philosophy was both endorsed and revamped as Larry Wilson shifted the scope and scale of Rohm and Haas, to prepare it for a new century in which global reach and expanding electronics would be key themes.[48]

The 1999 acquisition of Morton International provided Rohm and Haas with access to the profitable salt business. The Morton Salt box emblazoned with the image of the umbrella girl, below, is an advertising icon recognized by millions of consumers. The Morton umbrella girl first appeared on table-salt packages in 1914. Left: Fidencio Razo, mill leadman at Morton's Glendale, Arizona, solar salt facility, checks a sample of water conditioning salt for impurities, circa 1991.

ROHM
罗门哈斯中国研发
China Research and Development

08 A GLOBAL ENTERPRISE

One-World Thinking

Larry Wilson and Raj Gupta, who succeeded Wilson as CEO in 1999, pushed Rohm and Haas through the global door. That door had been opened by previous generations. Thinking about overseas opportunities was a deeply rooted tradition in the company. In the earliest days Otto Haas himself conversed and traveled quite naturally in both the Old World and the New. Following this lead, Don Murphy's foreign operations at first focused on Europe with a secondary emphasis on Latin America. Murphy and his international crew, along with successor Don Felley, also had the foresight to recognize the potential of the newly industrializing coun-tries—and sowed seeds there. In due course Murphy put up plants in India. Nearby opportunities were also on his radar screen, with sales offices opened in Japan, New Zealand, and Australia. India, a former British colony and natural market for

ROHM AND HAAS'S STRONG INTERNATIONAL GROWTH WAS FORESHADOWED BY A STRING OF MODEST BUT FORTUITOUS INVESTMENTS IN ASIA BY DON MURPHY. ONE SUCH ENTERPRISE WAS INDOFIL, AN AGRICULTURAL CHEMICALS PLANT IN INDIA, PICTURED HERE UNDER CONSTRUCTION IN 1965. PREVIOUS SPREAD: IN OCTOBER 2006 ROHM AND HAAS OPENED THE CHINA RESEARCH AND DEVELOPMENT CENTER (CRDC) IN SHANGHAI. WITH THIS FACILITY, SERVICES TO CHINA—R&D, SALES, MARKETING, AND CUSTOMER SERVICE—BECAME CENTRALIZED UNDER ONE ROOF AND MANAGED BY THE ASIA-PACIFIC LEADERSHIP TEAM.

Don Murphy's successor as head of the international department, Donald Felley, right, had cut his teeth in foreign operations and knew how to evaluate overseas investments.

U.K.-based multinationals, had passed laws mandating that foreign investors partner with local capitalists. These laws led Rohm and Haas to two equity joint ventures: Indofil Chemicals for agricultural products and Modipon for synthetic fibers. Formed in 1965, Modipon was running one of India's few high-tech nylon facilities within a few short years. Vince Gregory also urged the foreign-operations department to explore investment opportunities elsewhere in East Asia. For instance, in 1972, Ed Stanley, the company's Far East representative, toured Indonesia to study the market.[1]

Opportunistic and entrepreneurial, the international division also reached out to new markets in Eastern Europe, as it opened to U.S. business following the détente negotiated by American and Soviet leaders. Rohm and Haas's sales in Poland expanded as the country's chemical industry grew by more than 80 percent from 1971 to 1975. Using foreign exchange from exports, Polish industries purchased resins, oil additives, and a variety of finishes and emulsions for textiles, paints, and leather. James M. Hefty, the sales manager in Poland, worked closely with customers like Mazovian Refinery and Chemical Works and Gamrat-Erg, a PVC manufacturer. After talking with Gamrat-Erg's engineers about an injection-molding logjam, Hefty wrote a report that made its way to the Bristol plastics lab and caught the attention of Polish-American scientist Gene Szamborski, an expert in plastics intermediates. Recalling a similar problem in a U.S. plant, Szamborski created a new plastics modifier for the Gamrat-Erg works, whose engineers took delight in getting a "Polish-American solution to this Polish problem."[2]

Détente also opened the doors for joint ventures behind the Iron Curtain. Since the 1950s the European office had been doing business in Yugoslavia through the foreign-trade office. The agency slowly came to realize the advantages of making chemicals at home. In 1972 Rohm and Haas formed a joint venture with this Yugoslavian agency, creating Yugocryl to build an emulsion plant in Ljubljana, the capital of Slovenia. Customers included Industrija Usnja Vrhnika, a large leather manufacturer whose products were sold by boutiques and department stores in Western Europe and North America, and Dekorativna Ljubljana, a large textile mill that wove upholstery fabric for the Danish furniture industry. Another major customer was Jub, a paint manufacturer that became the third partner in Yugocryl. "We saw that the time for acrylics had come," and "by working closely with Rohm and Haas we could greatly facilitate our move into this new area," explained Jub director Florijan Regovee. "With the establishment of Yugocryl, Rohm and Haas Company became the first U.S. chemical producer to have a manufacturing operation in a Communist nation," Felley reported. "It represents a significant milestone for Rohm and Haas and an indication of our company's strong commitment to serving the world's markets."[3] Innovation through collaboration worked in Eastern as well as in Western Europe!

As sales in Eastern Europe grew, the international department kept a watchful eye on local industries like this Polish textile plant, shown at left in 1974.

A significant opportunity emerged in the Soviet Union, where the State Committee on Science and Technology signed technical agreements with thirty-eight American multinationals between 1972 and 1975. Agreements with Occidental Petroleum, General Electric, Monsanto, General Dynamics, Hewlett-Packard, Kaiser Steel, Coca-Cola, and Gulf Oil called for scientific cooperation in a number of fields, including power transmission, shipbuilding, oil production, and agriculture. Rohm and Haas had been doing business with the Soviets through its Austrian office since the early 1960s, selling products on contract to the state bureaucracy. The new arrangement promised to facilitate direct contact with manufacturers and farmers, permitting Rohm and Haas to do what it did best: "to adapt and formulate products specifically for given manufacturing processes and end uses."[4]

Before 1975 the company was managed geographically, with the U.S.-based divisions focused on chemicals, plastics, fibers, health, and licensing. The international division, which was highly profitable, sold a wide range of chemicals and was managed as three geographic segments: Europe, Latin America, and the Pacific. Each overseas office did its own thing, selling particular products suited to local markets without knowing or caring too much about how those sales or product lines fit into the larger corporate picture. This modus operandi was rooted in Murphy's entrepreneurial style and the era's limited communications technologies.

By the mid-1970s managers were recognizing the need to "look at things on a more worldwide basis." Vince Gregory's introduction of matrix management sat well with the need to think more strategically about the growing, changing overseas operations. Those operations lent themselves easily to the new matrix idea of a dual reporting structure based on products and geography. Four business teams—agricultural chemicals; plastics; polymers, resins, and monomers; and industrial

ROHM AND HAAS COMPANY | **TODAY**

DELIVERING RESULTS

The final link in the supply chain of Rohm and Haas products to its customers is not the shipment of materials: the company will continue to work with the customer on improvements. Here, warehouse operator Gerry Kurtz stages drums of finished product for shipment.

By the start of the 1980s Rohm and Haas was very much an international specialty chemical company. Innovation through collaboration with customers, along with an awareness of shifting markets, changing technologies, and competing geographies, defined its style. The world had become a fountain of wealth—and a source of creative worry.

chemicals—would oversee the products themselves, while distinct regional teams managed operations in North America, Europe, Latin America, and the Pacific. The business directors were responsible for strategic decisions, while the regional managers oversaw short-range goals.[5] Larry Wilson, who worked out of London from 1974 to 1977, summed up this strategic shift in his later comment that "starting in the mid-1970s, and I would say progressively every year, we have become more one-world thinking."[6]

Born and raised in India, Raj Gupta, pictured below (center) on a visit to Rohm and Haas Philippines, understood the advantages of investing in Asia and nudged fellow executives in that direction.

In 1977 North American sales still accounted for the majority of profits—almost $50 million after taxes—but the other three regions, Europe, Latin America, and the Pacific, contributed another $30 million. Europe led the pack at $13 million, followed by Latin America at $11 million and the Pacific at $5 million. Nearly 40 percent of the profits were coming from outside North America, a fitting testimony to a firm learning "one-world thinking."[7]

Creative Worry

By the start of the 1980s Rohm and Haas was very much an international specialty chemical company. Innovation through collaboration with customers, along with an awareness of shifting markets, changing technologies, and competing geographies, defined its style. The world had become a fountain of wealth—and a source of creative worry. If the 1970s had been a turning point in corporate thinking about the wider world, the 1980s was a time for putting these new ideas into action—and giving a younger generation the opportunity to find their way around the globe. Gregory asked the London office to identify business openings in Western Europe and sent experienced managers to the Pacific and Latin American regions "to pursue and explore investment opportunities."[8] Strategic planning shifted into high gear as Gregory pushed the international managers to make their subsidiaries more profitable.

Bright youngsters drawn from home operations to the international team included Rajiv L. Gupta and J. Michael "Mike" Fitzpatrick, both of whom would climb

Mike Fitzpatrick worked for Rohm and Haas in Brazil, Italy, Mexico, and England, combining his expertise in chemistry and in marketing to offer customer support. Above: In 2000 Fitzpatrick (left), then president of Rohm and Haas, met with the vice mayor of Shanghai City, China.

Yankees Abroad

J. Michael Fitzpatrick and his spouse, Jean, a Ph.D. in anatomy from Thomas Jefferson University, broke new ground at Rohm and Haas. In the spring of 1980 Mike was working as an agricultural marketer when John F. Talucci brought him down to the Latin America regional headquarters in Coral Gables, Florida. Talucci, looking for advice on launching Goal herbicide in Brazil, quickly realized that Mike's combined expertise in chemistry and marketing made him the ideal person to promote herbicides from the São Paulo regional office.

When Mike transferred to Brazil, dual career moves were uncommon at most American companies. Jean resigned her position as a Spring House toxicologist and accompanied her husband as a "trailing spouse." It soon became clear that "Jean was virtually the only toxicologist in Brazil," and so she was soon hired back by the company and set to work. In 1985 the Fitzpatricks moved again, this time to Milan, where both Mike and Jean had jobs waiting for them. He oversaw the agricultural chemicals business for Italy, the Middle East, the Balkans, and eastern Africa, while she worked as regulatory manager for Italian and Austrian sales areas. They both enjoyed the extensive business travel, frequently heading in different directions. On rare occasions they would bump into each other in a European airport, as their paths crossed on their different assignments. Jean was the company's first female expatriate, and the Fitzpatricks were its first double-expatriate family.

In Milan, Mike developed an understanding of the non-Western business environment, including that of Africa. He frequently traveled from Cairo to Johannesburg to visit customers and help local agents. In the Sudan any deal involved complicated logistics and a substantial commitment of manpower and equipment to customer support. "We weren't just *selling* the herbicides," Mike said. "We had to *apply* them as well." This state of affairs entailed storing the products in a Sudanese port, waiting for the payment to be deposited in the bank, shipping the product by truck to the state of El Gezira—a fertile agricultural district outside of Khartoum—and spraying the cotton fields by plane.

The jobs in Brazil, Italy, Mexico, and England exposed him and Jean "to lots of different business environments and lots of business practices." The couple became fluent in Portuguese, Spanish, and Italian, made enduring friendships with local people, and soaked up the vibrant commercial scene in Milan. "I was lucky enough to be involved in a very diverse commercial environment almost everywhere I worked," Mike explained. "I worked through hyperinflation in Brazil. I negotiated with the heads of Communist cooperatives in Yugoslavia. The experiences built on each other." And, as with other leaders, the experiences prepared him for his eventual move in 1999 to become president and chief operating officer of Rohm and Haas.

Condensate Polishing
Your Lineup for the 80's

Ion-exchange resins became a legacy business that Rohm and Haas repeatedly reinvented to keep up with changing times. Above, researchers analyze Amberlite resins at the laboratory in Chauny, France. Right: Amberlite resins were promoted to power stations with this ad in Chemical Engineering, Power Engineering, *and* Power *during the early 1980s. When a power station releases steam to generate electricity, the cooling steam condenses back to water, having picked up impurities along the way. To remove the contaminants, the water is passed through a condensate polishing unit filled with ion-exchange resins.*

the corporate ladder to become, respectively, CEO and chief operating officer in 1999.[9] Born in Muzaffarnagar, India, and a graduate of the prestigious Indian Institute of Technology in Mumbai and of Cornell University, Gupta arrived at Rohm and Haas in 1971 from nearby Scott Paper, fresh with an M.B.A. from Philadelphia's Drexel University. Thinking he might eventually return to his homeland, Gupta accepted a job at Rohm and Haas because it had subsidiaries in India—but he never ended up at one of them. Starting out in finance, Gupta soon transferred to international operations, managing English and French businesses for more than a dozen years before assuming broader responsibilities for the Asia-Pacific region. Fitzpatrick, in his turn, joined Rohm and Haas from a National Institutes of Health postdoctoral fellowship at Harvard University, devoting five years to agricultural research at Spring House before taking a series of marketing and management jobs in Latin America and then Europe.[10] Both men's career experiences show how Rohm and Haas built on the collaborative practices pioneered by Otto Haas and Don Murphy and adapted them to the new global economy.

Gupta earned his stripes in London and Paris, running different aspects of the European business as he inched toward general management. Although apprehensive about leaving the United States for the United Kingdom—largely because of the history of England-India relations—he set aside his worries and accepted the English assignments for the chance to try something new. Later in France, Gupta mastered an unfamiliar language and learned new skills in sales, marketing, and customer relations—and started to appreciate how all the functions could be woven into a cohesive whole.[11]

In 1984 he became business manager of the European ion-exchange business, recommended for the position by Basil Vassiliou, himself poised to become regional director out of London. The job put Raj Gupta in the hot seat, engineering the acquisition of Duolite International, a rival French manufacturer. Starting in the late 1970s Rohm and Haas lost money on ion exchange, by now a legacy business, mainly owing to competition from Duolite, which was owned by the U.S. firm Diamond Shamrock. Vassiliou decided to buy Duolite—the $60 million purchase would be Rohm and Haas's biggest acquisition to date—but market regulators in Brussels refused to approve the merger. If that were not enough, the French employees threatened to strike. Vassiliou convinced French authorities that the merger would make France into Europe's largest ion-exchange manufacturer, securing their help with European Union officials. At the Paris office the young Raj faced the formidable challenge of completing the integration after the top Duolite people quit. "This was my first general management experience," he remembered, "and I barely spoke any French." Gupta's solution for peace and productivity was twofold: assign people from both companies to key positions in the new subsidiary and refocus energy on positive goals like customer satisfaction.[12]

Over the course of the decade Gupta played a seminal role in restructuring the European business. The Duolite job went hand in hand with the strategic imperative of improving regional performance, as the European Union created a unified market. Years later Gupta described the challenge: "We recognized that we needed to become efficient in order to grow. We didn't cut; we consolidated. Instead of having three plants making Dithane—one in Italy, one in France, one in England—or having little product lines in each of the countries, we recognized that we were now operating in the European Union. Duties were gone, and we needed to have fewer facilities, so they could become efficient and source all of Europe. That's how we saved the European business." Gupta, together with Vassiliou and other one-world thinkers, transformed an aging business into a moneymaker.[13]

Eyes on the Future

In 1988 Gupta went back to London, rising to become the worldwide business director for plastics additives. "Rohm and Haas had adopted a global organization," he explained. The best way to run the business in a world of growing communications and competition was to organize by product line.[14] A significant further change to the matrix structure was introduced by Larry Wilson, who became CEO in 1988. Gupta was selected for the role of pioneer, becoming the first to manage a global unit from someplace other than Philadelphia. In the era before ubiquitous cell phones, e-mail, and ERP systems, "this was more than a subtle change. Running a global business away from the home office was a great experiment," Gupta remembered. "Today we don't give it a second thought. In those days no one knew if a global business could be run far away from the primary production site or the primary market or the center of corporate management. We demonstrated that it could be done."[15]

Basil Vassiliou had launched his Rohm and Haas career as an ion-exchange specialist and climbed the corporate ladder by helping build the European business. Above, Vassiliou (left) is photographed with Kurt Petry, manager of Rohm and Haas Deutschland, at a dedication ceremony in 1987.

As he transformed his $300 million unit from a product-driven to a market-oriented business, Raj Gupta also initiated a Japanese alliance that introduced a new way of doing business globally. A long-standing relationship with the Kureha Chemical Industry Company needed an infusion of energy and trust. Gupta responded by designing a win-win solution. The Japanese maker of plastics additives was providing the know-how but did not have a stake in the Rohm and Haas subsidiary that used their technology. Gupta's solution entailed giving Kureha part ownership of a plant in Grangemouth, Scotland, in exchange for an interest in its new manufacturing facility in Singapore. Each company continued to run its subsidiary, but the boards were diversified with directors from the partner firms. "This type of sharing," Gupta explained, "really made our alliance global." The Kureha deal was significant in another way: "It was clear that Asia was going to become important, so Rohm and Haas wanted a foothold there." The joint venture lasted from 1988 until a 2003 buyout by Rohm and Haas and provided entry to Asia's plastics market.[16]

CEO Larry Wilson worked with a trusted team of executives to look for opportunities to expand Rohm and Haas's global reach. Above, Wilson is flanked by the next generation of leaders, Mike Fitzpatrick and Raj Gupta, on the cover of Chemical Week, *February 17, 1999. Gupta, shown opposite circa 2008, became Wilson's vice president for the Asia-Pacific region in 1993.*

Looking to the company's future, Wilson summoned Gupta back to Independence Mall in 1993 to become a vice president. Returning to Philadelphia did not sound like a promotion, given the company's history of developing executives internationally, but the boss allayed his fears. Since the mid-1980s Rohm and Haas had become more profitable, but it was not growing. Gupta's levelheaded approach to risk assessment made him the perfect vice president for the Asia-Pacific region, with its challenging but rapidly growing markets.[17]

Wide, Wide World

In the summer of 1995 Larry Wilson gathered his trusted advisers for a two-day meeting in the ninth-floor boardroom at the home office. This close-knit group of Philadelphia-based executives—a coterie of vice presidents together with the chief financial officer—helped Wilson, the CEO and chairman, think through strategic matters. In keeping with tradition the meeting was to focus on a pressing issue. The issue chosen arose from Wilson's familiarity with the wide, wide world—and his understanding of its importance to Rohm and Haas. As a young man Larry Wilson had honed his management skills as regional director for Europe, running the $150 million operation. Now, acutely aware how rapidly the business relationships between the United States and other nations were changing, there was a question he could not ignore. How should the company expand its presence in the emerging markets of the Asia-Pacific region, and, most important, how should it view the rising Red Dragon, China?[18]

Rohm and Haas already had a very modest interest in China. By the late 1980s the company had a joint venture in marketing with acrylic maker Beijing Eastern. More recently it had agreed to collaborate with the Shenyang Research Institute, the agrochemicals arm of the Ministry of Chemical Industry, to commercialize new herbicides, fungicides, and insecticides. Now Wilson mulled over the decision whether or not to build three plants, significantly expanding the firm's presence in China. The $50 million expenditure was not large for a $4 billion company. However, the long-term impact of the decision could be of major consequence, given China's rapidly evolving marketplace and its rising middle class hungering for consumer goods—to say nothing of the accelerated pace at which other foreign companies were setting up operations in China.[19]

Wilson's team turned their full attention to Raj Gupta, who was arguing the case for China. Each year now, he was spending six to eight months in Asia, getting to know markets and customers. Initially Gupta traveled with predecessor Bill Kulik or with Larry Wilson himself. Soon he was on his own, living out of suitcases as he explored realities and opportunities within his vast territory.[20]

Gupta had befriended local managers like Mark McGuire in Thailand, George Presanis in Singapore, Ruben Salazer in Japan, and Tom Grehl in China. He learned that these people "really believed Rohm and Haas could and should do more in Asia" and were skeptical of the company's "stop-and-go" efforts. In two businesses—electronic chemicals and ion-exchange resins—the Asia-Pacific region was already the firm's largest contributor to sales and earnings. While new plants in Thailand, Indonesia, and Taiwan received ready approval, the home office seemed befuddled by China. "China was the big nut to crack," Gupta recalled. "Some of the projects for China were on the books for consideration by management for almost five years. . . . We used to talk about China. People would visit there and come back very excited, but nothing ever moved." In a fall 1994 meeting with his frustrated Asia-Pacific team Gupta promised to "take action with the executives in Philadelphia." Now he and Tom Grehl, who directed Rohm and Haas China, were advocating a big push.[21]

Gupta and Grehl reviewed the three extant Chinese businesses—emulsions, agricultural chemicals, and ion-exchange resins—and explained threats, opportunities, and risks to the Philadelphia executives. Initially "everyone agreed that China was going to be an important country," Gupta remembered. Then on the second day the mood changed. Concerns were raised about China's approach to intellectual property.

During the 1990s Rohm and Haas's expansion efforts shifted from European and Latin American markets, as featured in the Autumn 1984 Reporter, *right, to Asia. In 1991 Rohm and Haas opened a Beijing office to provide better service to customers in the Chinese leather industry. The staff is pictured above in Tiananmen Square (left to right): Jane Li, Lincoln Li, Tom Grehl, and Jason Zhang.*

Someone wondered if it would be possible to maintain safety standards for ion-exchange production in a Chinese facility. One by one, experienced senior executives posed tough questions and established reasonable doubt until "it felt like China projects would be dead in the water." When Gupta tried to reignite the fire, the discussion turned back to the risks—until Wilson spoke up. "Time out. We're going to do all three of the projects." As Gupta himself puts it, "Larry Wilson was the man who had the foresight to say it was time for Rohm and Haas to put a larger stake in Asia. He had this intuition about Asia."[22]

Wilson's giving Gupta the freedom to learn about the region, and his decision to expand manufacturing in China, pushed Rohm and Haas into a new era. To be sure, Wilson surrounded himself with seasoned managers familiar with foreign investments. "One unusual part of Rohm and Haas," he has said, is that "virtually the entire operating management comes out of the international side of the business." Basil Vassiliou grew up in Greece, earned his doctorate at the Technical University of Athens, and spent decades in Europe, having founded the R&D center at Valbonne near the French Riviera. Don Garaventi was a veteran of Latin America, where he had run businesses in Colombia, Argentina, and Mexico before managing the Asia-Pacific region in the late 1970s.[23] In a later era Pierre Brondeau, Raj Gupta, and Mike Fitzpatrick provide additional examples. Savvy executives like these knew how to assess opportunities outside the United States, but everyone was cautious when it came to China, the great unknown.

Managers like Tom Grehl, a Columbia M.B.A. fluent in Mandarin, Cantonese, and Japanese, were on the sidelines until the Wilson-Gupta eras, when their familiarity with Eastern cultures gained wider recognition in the company. Grehl had seen the Chinese people bend over backward to help a supplier or customer. "In China, business is not business," he liked to say. "Business is friendship."[24] Wilson and Gupta were of like minds, knowing the firm had to support the judgment of local people like Grehl if Rohm and Haas were to succeed in Asia. "We went ahead and built all three of the plants in China," Gupta explained, thanks to "Larry's willingness to make a top-down decision based on instinct, gut, and confidence in the people on the ground who would do their very best to make it work."[25]

Eventually Wilson was also to steer the company through some heart-wrenching divestitures, including Plexiglas, the acrylic that was a household name. Shedding commodities like Plexiglas caused the company to shrink in ways that could threaten its viability in the long term. Counterbalancing opportunities for growth were a top priority. And Asia offered a compelling case, as the increasing Rohm and Haas commitment to electronic chemicals meshed with Asia's burgeoning role in electronics. Wilson and Gupta kept their fingers firmly on the pulse, as major parts of the electronics industry moved to the region.

Two key acquisitions—Rodel and LeaRonal—augmented Rohm and Haas capabilities in electronic materials. A 1997 alliance with Rodel, a Delaware firm that invented a chemical for polishing semiconductor surfaces, eventually led to a merger. The 1999 acquisition of LeaRonal, which like Shipley had established Far East operations back in the 1970s, further strengthened Rohm and Haas's Asian presence and added a series of new products for printed-circuit-board manufacturing, semiconductor packaging, and electronic-connector plating. In the promising field of electronic chemicals the company's major customers now included Samsung, Sharp, Intel, Lucent, AMD, IBM, and Phillips.[26]

Smart Globalization

When Wilson passed the mantle to Gupta in 1999, Rohm and Haas was organized on a global basis and had become the world's largest manufacturer of specialty chemicals, with major commitments to electronics and to Asia. Raj Gupta became the fifth chief executive at Rohm and Haas, as the world economy was still reeling from the Asian financial crisis of 1997–98. By early 2000 it was apparent that an economic downturn would hit the United States. Construction slowed, the cost of raw materials rose, and the dollar weakened against European currencies. Gupta, who had spent more than a dozen years grappling with the ups and downs of the European business, remained undaunted. Building on Wilson's work, he would spend the next decade expanding strengths in chemical electronics, pushing deeper into East Asian markets, and developing opportunities elsewhere around the globe.[27]

Among those he relied on was Pierre R. Brondeau. Characteristic of the new breed of global-oriented managers, Brondeau, a Frenchman, had a Ph.D. in biochemical engineering from the Institut National

As Raj Gupta took over the reins from Larry Wilson, he relied on a new breed of global-oriented managers like Pierre Brondeau. Above: Brondeau, president and chief operating officer, listens to a presentation at the home office in 2008.

des Sciences Appliquées de Lyon and significant industrial experience when he joined Rohm and Haas in 1989 as European marketing manager for the plastics additives business in Paris. Brondeau soon transferred to the Valbonne R&D center as head of technical services, and by 1995 he was director of research, sales, and marketing at the Shipley Company. From there it was a short step to his 1999 appointment as CEO of Shipley and vice president and business group director of the electronic materials division.[28]

Gupta and Brondeau realized that while the United States and Western Europe were still viable markets, their growth was paling in comparison to that in Asia. Mature economies were expanding at a mere 2 to 3 percent per year compared with 5 to 10 percent for emerging markets. China was becoming the new "workshop of the world," while India had positioned itself as a skilled service provider, a hub for global

ARRAY OF USES

Rohm and Haas has developed dispersants that are used in a range of applications, from dish and laundry detergents to paint and coatings to industrial mineral processing. Here, senior applications scientist Somil Mehta runs an experiment on dispersants.

ROHM AND HAAS COMPANY | **TODAY**

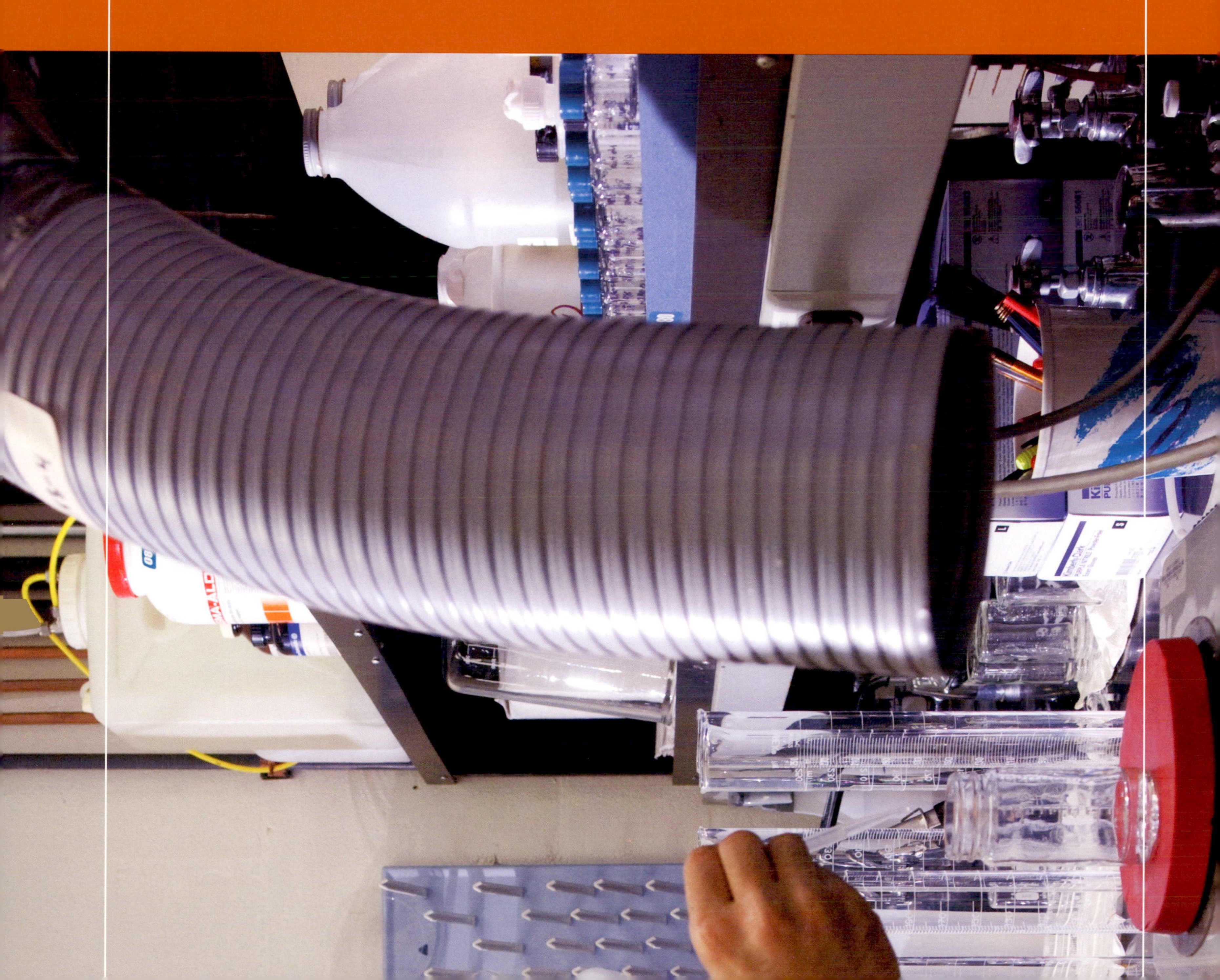

Raj Gupta: Numbers Guy

When Rajiv L. Gupta had been at Rohm and Haas for five years, he started to wonder whether his talent for financial analysis would lead his bosses to categorize him as a "numbers guy." Gupta asked for some overseas experience to help him move into general management. When the financial director for the U.K. operation became ill, Vince Gregory and Fred Shaffer arranged for Gupta to take his position. So began a fifteen-year stint in Europe, where Gupta took a series of jobs, spending five years in London, five in Paris, and five again in London.

"The idea of moving from the United States to England wasn't my idea of a good overseas assignment, given the history of English-Indian relations," Gupta remembered. The first six months in London "were a bit of a transition." But in the long run "England turned out to be a great assignment."

Gupta witnessed the impact of Margaret Thatcher's policies, which helped modernize the economy. "When we arrived in the U.K. during the late 1970s, England was truly still an island. The towns had tiny stores, which closed at five o'clock during the week and were only open half a day on Saturday. Five years later we found a huge change. The Conservative Party had dealt with the labor issues, and big stores had arrived. Suddenly, you saw food from all over the world in the supermarkets. England had reemerged after a period of post-industrial decline, and the Conservative Party had dealt with the legacy of old issues."

Before the era of globalization and multiculturalism Gupta was an atypical example of an executive representing an American company in Europe. "People seemed curious about how someone like me ended up in Europe," he remembered. In London during the Brixton riots of 1981, Gupta knew that racial tensions marred English social relations, but he found few barriers at the office. "At work relationships were seamless. The great thing about Rohm and Haas is that, no matter where you go, people are polite, respectful, and treat their colleagues in the same way, regardless of their background."

At Gatwick Airport, Gupta once stood in a three-hour line at passport control. "There was a line for U.K. citizens, and a long line for 'all the rest.' With an American passport I got stuck in the long line for 'all the rest.' When I got to the immigration officer, he looked at the passport and said, 'You were born in India, you lived in France, you're living in England, and you're an American citizen. How did you do this?'" Gupta responded: "By paying a lot of taxes to Her Majesty's government."

Raj Gupta, like Otto Haas, was an immigrant who adapted to his new American circumstances, always keeping his eye on the wider world and the bigger picture. "I want one of my legacies to be helping people firmly believe that Rohm and Haas is a company that wants to do things the right way and deliver superior results," Gupta, above, said in 2007. "Innovation has been the history of this company, and it will be its future."

Although consumers were not directly aware of many of its products, Rohm and Haas sat in the middle of a supply chain that began with the raw materials produced by large oil companies . . . and ended with household goods like paints, cosmetics, and electronics sold by big-box retailers, including Home Depot and Wal-Mart. These latter giants, armed with price-setting power, squeezed the profits of suppliers like Rohm and Haas.

outsourcing in information technology. These megatrends were driving fundamental changes in the chemical industry, which operates in the shadow of manufacturing and services. Rohm and Haas necessarily followed its customers in the construction and electronics industries, who in turn were moving in response to their customers. "It is not labor costs that drive change," Gupta explained to *European Chemical News*. "It's proximity to customers and raw materials." Corporate decision makers who wanted to act smartly identified the shifts early and figured out how to make the best of them. However, before Gupta could embark on his program of "smart globalization," he had to wrestle with the troubled economic environment.[29]

The new CEO faced the arduous task of restructuring Rohm and Haas in parallel with the broader pattern of consolidation in the chemical industry. It was clear that the company had too many manufacturing sites, mandating a reassessment of resources. Although consumers were not directly aware of many of its products, Rohm and Haas sat in the middle of a supply chain that began with the raw materials produced by large oil companies like BP and ExxonMobil and ended with household goods like paints, cosmetics, and electronics sold by big-box retailers, including Home Depot and Wal-Mart. These latter giants, armed with price-setting power, squeezed the profits of suppliers like Rohm and Haas. At the same time, American corporations faced a slew of expenses, such as higher pension-fund investments, post–9/11 outlays for security, health-care costs, and compliance with Sarbanes-Oxley legislation, which set higher standards for corporate ethics. "I believe that differentiating products is the only way to succeed," Gupta told the Société de Chimie Industrielle, an executive group, in 2003. Rohm and Haas had to stay focused on its customers and on innovating with them, thinking more creatively about the ultimate consumer while finding ways to reduce costs.[30]

Gupta charted a course through the storm, combining proven tactics with newer ideas. He cut jobs, repositioned assets, and ramped up efforts to bring new products to market. One bright spot was the continued growth of electronics under Brondeau's leadership. Electronics was also important in its role as an in-house

The rising standard of living in India provided opportunities to Rohm and Haas, which opened a $20 million manufacturing facility in Taloja, near Mumbai, in 2003. The plant made adhesives, sealants, and coatings and provided technical service to the customers. Below, a worker runs viscosity tests in the laboratory, circa 2004.

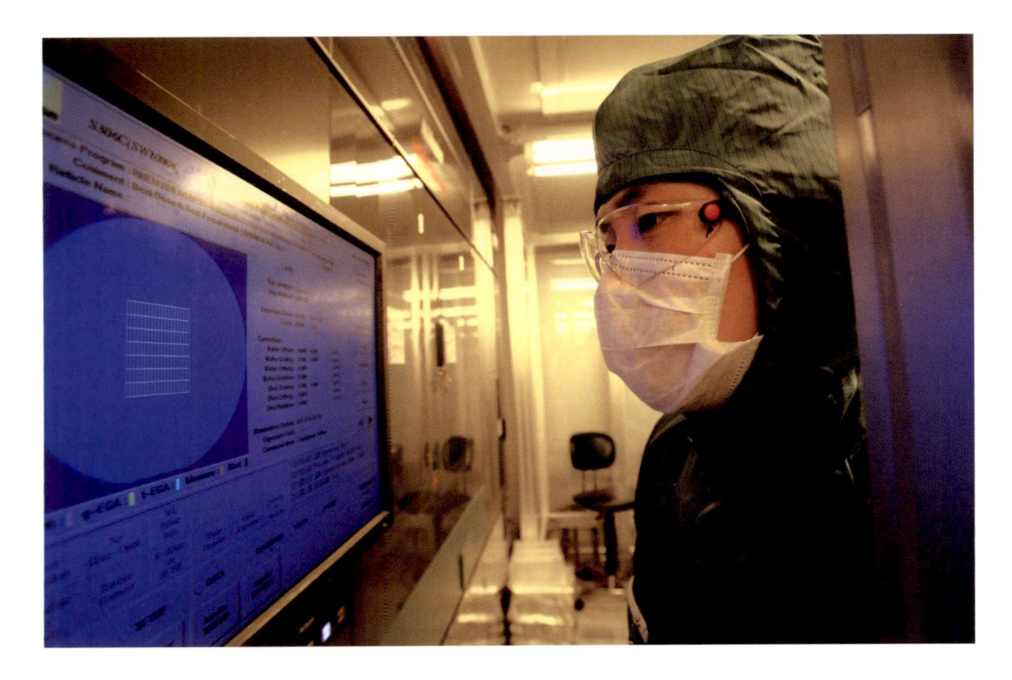

Rohm and Haas continues to bring best practices to all of its overseas operations. Precision instruments like the Nikon Stepper in Cheonan, Korea, above, are important tools for achieving the quality required in the microelectronics industry. Opposite: At Rohm and Haas's Croydon plant in Pennsylvania, operator Steve Zluky reviews a process with Murgesh Konar of the Rohm and Haas facility in Taloja, India, as part of the latter's training process.

technology, as well as being a product line. As electronic forms of communication multiplied while simultaneously making it seem that the world is flat by eliminating geographic barriers to competition, the firm struggled to install a $300 million global ERP software platform designed to facilitate quicker response to the marketplace. Older information systems combined with disparate technologies inherited by merger threatened chaos. In some instances frustrated managers gave up on the intranet, using fax machines to take orders and only then entering them into the computer. Connecting offices from São Paulo to Shanghai, the global ERP system fostered better communication, improved response time with customers, and enabled one-world thinking.[31]

Life and business had changed dramatically in the ninety years since the two Ottos had partnered to sell Oropon bate to local tanneries. Otto Röhm's chemical prowess was legendary, and Otto Haas's personal, hands-on approach still shaped customer service. But the world had grown smaller and people were drawn closer together, thanks to technologies like long-distance jets, laptop computers, and cell phones. Managers could no longer rely solely on the genius chemist to invent products that salesmen could push. In 2002 Gupta described the new world in a keynote at the American Chemical Society: "Technology and innovation remain the essential drivers of success for a specialty chemical company. There is still a wealth of great ideas emanating from chemists at the lab bench, and this research needs to be supported with strong R&D spending. With the wide-open ability of science to hopscotch across every element in the periodic table, incredible computing power at our fingertips, and the connectivity of the Internet, I believe there are almost limitless opportunities for innovation. And that's the problem. The challenge today is not finding the single 'eureka' technology to bring to the marketplace, but in making clear choices among limitless new product opportunities, based on knowledge of the marketplace and the consumer preferences that shape them."[32]

The challenge was to choose carefully, concentrating on a few technologies and developing specialized expertise around them. In the face of megatrends like globalization and consolidation the path to market differentiation required "a multifaceted approach." With five global divisions—coatings, electronic materials, adhesives and sealants, performance chemicals, and salt—Rohm and Haas focused on product areas that could command high premiums, such as acrylics and nanotechnology. Gupta had learned the perils of undisciplined diversification when, as a young financial analyst, he had watched his mentors deal with fibers and pharmaceuticals. He also recognized that it was foolish to rely exclusively on homegrown R&D. With smart globalization it was necessary to "cast a broad net," much as he did years earlier with the Kureha alliance in plastics additives. Raj Gupta's approach to innovation included new mechanisms for acquiring and developing products: alliances, joint ventures, intellectual property purchases, academic research, and marketing partnerships.[33]

Globalization introduced another variable: the imperative to accommodate a wide variety of local needs. Gupta expressed it in 2006: "The idea of standardized global products is passé." In China, where sales grew from under $15 million in 1993 to

ALL RAW MATERIALS

In 2006 Raj Gupta launched Vision 2010, a plan to position Rohm and Haas to be more competitive in emerging economies. Right, Gupta speaks at the 2006 grand opening of the China Research and Development Center in Shanghai.

$150 million in 2002, the Paint Quality Institute established a branch to publicize water-based coatings and, along the way, to understand what made Chinese households distinctive. In dealing with Chinese customers Rohm and Haas provided traditional technical support along with advice on how to comply with the environmental regulations introduced after China gained admission to the World Trade Organization in 2001. In India, Rohm and Haas sales totaled a mere $14 million in 2002, but there was great potential, given the size of the population and the similarities between the Indian, European, and American business models. Indian customers appreciated hands-on technical assistance, which gave Rohm and Haas an edge over lower-cost competitors. Harish Badami, managing director of Rohm and Haas India, explained how the customer-service tradition had been expanded to include the ultimate consumer. "We act as a consultants to our customers in the markets. We often help them reformulate a better product. Rohm and Haas has the ability to understand not only its immediate market but also its customer's market."[34]

Vision 2010

Reorganization is both a managerial prerogative and a recurrent necessity in a rapidly changing world. Gupta and his management team of Jacques Croisetière, Pierre Brondeau, and Alan Barton had accomplished much in the first half-dozen years since taking command orchestrating major portfolio moves, overseeing the integration of forty acquisitions, and dealing with harsh blows to the economy. The Asia footprint was deeper, the commitments to China, India, and other developing economies stronger. Balance sheets showed some of the best profit margins in the company's history. This payoff provided Gupta with the launch pad for a major realignment, announced in 2006 as Vision 2010 and designed to position Rohm and Haas to capitalize on further changes in the global environment.[35]

Vision 2010 introduced a business structure focused on three core market areas—electronic, specialty, and performance materials—and an operational model focused on becoming more competitive in emerging economies. Around 20 percent of the company's sales now came from the recently industrialized nations, and Gupta hoped to push the figure up to 35 percent. Much of the growth would come from the coatings and electronics businesses in key markets of the Asia-Pacific region.[36] Within electronics there would be a whole new platform in display technologies.

In the forward plan the roaring Red Dragon had company: the Russian bear and the Polish eagle. Back in the mid-1990s Gupta and his colleagues recognized that Eastern Europe, with its well-trained workforce, industrial base, and curiosity about capitalism, would be the new frontier. Although some U.S. and European investors considered the shaky political climate a liability, Gupta told *Chemical Marketing Reporter* that the region would rise. By 2006, to his satisfaction, Rohm and Haas forecasts projected vigorous growth in emerging markets in Asia—and central and Eastern

Europe.[37] Rapidly developing economies like Russia and Turkey were now on the radar screen.

Vision 2010 updated the global organization put in place during the late 1980s. Global units had served Rohm and Haas well, teaching managers in the field to appreciate the importance of regional customs, needs, and desires. Customers in particular regions understood their own local markets and demanded appropriately customized products. Construction techniques, transportation systems, and beauty products varied from locale to locale. Consumers in Brazil and Bangkok had different expectations. "In order for us to rapidly respond to customer needs for products and services," Gupta explained, "we must rebalance our structure toward decentralization."[38] In some areas—finance, human resources, information technology, procurement, safety, and strategy—global standards still made sense. At the same time, Rohm and Haas aimed to become more self-contained in each region, focusing on particular customer needs, taking advantage of neighborhood prices for raw materials, and cultivating local talent familiar with the language and customs.

In 2008 Pierre Brondeau—a practiced hand in global, specialty business—was named president and chief operating officer. This recognition of his day-to-day leadership of electronic materials through an era of rapid growth itself set the stage for the tasks ahead. In the new global marketplace of complexity, change, and customization, the ability to cater to local differences mattered more than standardization. The specialty chemical manufacturer who could offer something truly special would win the day.

Rohm and Haas at 100

As the global economy boomed in the years leading up to 2009, Rohm and Haas was nimble and adaptive. Its focus on electronic, specialty, and performance materials mirrored the ways in which new chemical knowledge connected to buoyant, growing markets. In this, the era carried its echoes from the years around 1909, when Otto Röhm's knowledge of bating made Oropon the chemical of choice for high-fashion vici leather. Nor was the era different in the ways it rewarded the firm that understood the needs of its customers and worked with them. What was different was the scope and scale of the operations of Rohm and Haas. The modest partnership of 1909 was by 2009 the world's leading specialty chemical company.

With over one hundred facilities in twenty-seven countries Rohm and Haas chemicals are deployed across the economy and around the globe. Its people and products are found in areas from building and construction, to food, packaging, paper, and personal care, to electronics and transportation. Otto Haas, his colleagues, and his successors have truly contributed to making the world a better place. The secret? Innovation through collaboration.

Employees at the Min-Hsiung plant in Taiwan, established in October 2006, celebrated the one-hundredth anniversary of Rohm and Haas in 2009. In the photo above, posted to a Rohm and Haas Web site dedicated to the anniversary, the Min-Hsiung employees conveyed good tidings to their fellow workers around the world.

Endnotes

Chapter 1

1 *The Haas Stories* (Philadelphia: privately printed, 1991), 58; Oropon notebook, entries for Robert H. Foerderer, Inc., 1909–12 Dec. 1913 ("troubles," 24 Oct. 1910), box 7, folder 8 [hereafter cited as 7/8], accession I, Rohm and Haas Archives, Philadelphia, PA [hereafter cited as ROH I].

2 John J. MacFarlane, *Manufacturing in Philadelphia, 1863–1912* (Philadelphia: Philadelphia Commercial Museum, 1912), 73 ("goatskins").

3 Joseph J. Stellmach, "The Commercial Success of Chrome Tanning," *Journal of the American Leather Chemists Association* 85 (Nov. 1990), 419–420.

4 *The Haas Stories*, 29; Sheldon Hochheiser, *Rohm and Haas: History of a Chemical Company* (Philadelphia: University of Pennsylvania Press, 1986), chap. 1; Rudolph Cronau, *German Achievements in America* (New York: Rudolph Cronau, 1916), 112.

5 F. O. Haas, "How Dr. Röhm and Mr. Haas Met," 21 April 1972, 34/33, ROH I; Bill Kohler, transcript of oral history with Sheldon Hochheiser, 1 July 1983, 1, ROH Archives [hereafter cited as Kohler transcript, ROH].

6 Charles P. Vaughn, *The Leather and Glazed Kid Industry in Philadelphia* (Philadelphia: Chamber of Commerce, 1917), 3; R. A. Church, "The British Leather Industry and Foreign Competition, 1870–1914," *Economic History Review*, new ser., 24 (Nov. 1971), 544.

7 *Twelfth Census of the United States, 1900* (Washington, DC: Government Printing Office, 1902), IX, "Manufacturers," pt. III, 733, cited by Church, "British Leather Industry and Foreign Competition," 544.

8 Vaughn, *Leather and Glazed Kid Industry in Philadelphia*, 3–4; *Nothing Takes the Place of Leather* (New York: American Sole and Belting Leather Tanners, 1924), 15–18.

9 Ernst Trommsdorff, *Dr. Otto Röhm: Chemiker under Unternehmer* (Dusseldorf: Econ Verlag, 1976), typed translation in ROH Archives, pt. I [hereafter cited as Trommsdorff translation].

10 Ibid.

11 Ibid.

12 William K. Andress to John Marshall, 16 Oct. 1905, 7/7, ROH I; Trommsdorff translation, pt. 1; Hochheiser, *Rohm and Haas*, 6.

13 Hochheiser, *Rohm and Haas*, 6–9.

14 "Otto Haas (1872–1960)," 34/33, ROH I; Donald S. Frederick, transcript of oral history with Sheldon Hochheiser, 27 Sept. 1983, 79, ROH [hereafter cited as Frederick transcript].

15 Harold Turley, transcript of interview, 27 May 1981, 35/17, ROH I [hereafter cited as Turley transcript]; Hochheiser, *Rohm and Haas*, 9.

16 Cronau, *German Achievements in America*, 112.

17 *Shoe-ology, How to Buy Shoes and How to Take Care of Them* (Philadelphia: Robert H. Foerderer, 1897), 9, Trade Catalog Collection, Hagley Museum and Library, Wilmington, DE [hereafter cited as HML].

18 Regina Lee Blaszczyk, *American Consumer Society, 1865–2005: From Hearth to HDTV* (Wheeling, IL: Harlan-Davidson, 2009), pt. I; Sears, Roebuck and Co., *Consumers Guide* (Chicago, IL, 1909), 926.

19 J. Edgar Rhoads, transcript of oral history with Norman B. Wilkinson, Lucius F. Ellsworth, and Faith K. Pizor, Greenville, DE, 1969, 70–72, acc. 2326, Manuscripts and Archives, HML [hereafter cited as Rhoads transcript]; Vaughn, *Leather and Glazed Kid Industry in Philadelphia*, 10.

20 C.M. [Carl Muckenhirn], "Oropon," in *The Oropon* 1 (April 1920), [8], 8/2, ROH I.

21 *Workshop of the World* (Wallingford, PA: Oliver Evans Press, 1990); Philip Scranton and Walter Licht, *Work Sights: Industrial Philadelphia, 1890–1950* (Philadelphia: Temple University Press, 1986).

22 *The Haas Stories*, 5, 8, 14. Wenzel worked for the company from 1913 until his death in 1945; see Rohm and Haas, Annual Report, 1945 [hereafter cited as ROH, 1945 AR], 64. All annual reports are in the Rohm and Haas Archives. Frederick transcript, 79.

23 Mira Wilkins, "The Neglected Intangible Assets: The Influence of the Trade Mark on the Rise of the Modern Corporation," *Business History* 34 (Jan. 1992), 66–95; idem, "German Chemical Firms in the United States from the Late 19th Century to Post-World War II," in *The German Chemical Industry in the Twentieth Century*, ed. John E. Lesch (Boston: Kluwer, 2000), 295.

24 Otto Haas to Charles Neave, 7 Dec. 1914, 76/4, ROH I.

25 *The Haas Stories*, 5–6, 24, 32; John C. Haas, conversation with Arnold Thackray, July 2008 (height).

26 Statement of Otto Haas, 21 May 1913, in *Otto Röhm, Individually, and Otto Röhm and Otto Haas, Co-Partners, Doing Business under the Firm Name and Style of Röhm and Haas v. The Martin Dennis Company*, U.S. District Court for the District of New Jersey, 16 July 1918, 49–67 (57, quotations), 77/1, ROH I.

27 "Otto Haas (1872–1960)." On the distinctiveness of tanning see "Testimony of Hon. Robert H. Foerderer," 21 Dec. 1900, in U.S. Industrial Commission, *Report of the Industrial Commission on the Relations and Conditions of Capital and Labor* (Washington, DC: Government Printing Office, 1901), 315–321.

28 Rhoads transcript, 73–74; Oropon notebook, entries for Kaufherr and Co., 1909–1916.

29 On Newark see Oropon notebook, entries for Berkowitz, Goldsmith and Spiegel, 8 June 1917–19 Mar. 1934 (quotations: 12 July 1920, "shut"; 14 Nov. 1920, "better"; 26 Apr. 1922, "Beaver").

30 Oropon notebook, entries for F. Blumenthal and Co., 1909–22 Nov. 1916 (quotation: 3 Dec. 1913).

31 Statement of Otto Haas, 57; Rohm and Haas, Darmstadt, to Haas, 19 Dec. 1913, 72/1, ROH I; Haas to Horace Pettit, 22 Dec. 1913, 5 Mar. 1914, 72/1, ROH I; statement of Charles Glen Beadenkopf, 10 May 1913, in *Röhm et al. v. The Martin Dennis Company*, 23–48.

32 "Third Annual Report of the U.S. Tariff Commission," *Shoe and Leather Reporter* (8 Jan. 1920), 96.

33 Kathryn Steen, "German Chemicals and American Politics, 1919–1922," in Lesch, ed., *German Chemical Industry in the Twentieth Century*, 326–328.

34 Hochheiser, *Rohm and Haas*, 23–25.

35 Charles Neave to Haas, 17 July and 16 Sept. 1918, 76/7, ROH I; "Report of Mr. Kelton," 21 Apr. 1920, 76/7, ROH I; A. Mitchell Palmer, advertisement, "To Be Sold by the Alien Property Custodian," *New York Times*, 13 Nov. 1918.

36 U.S. House Committee on Patents, *Investigation of Röhm and Haas Company Royalty Payments to Germany*, 77th Cong., 1st sess., 10 June 1941, 10.

37 ROH, 1947 AR, 16.

38 Haas to Van A. Wallin, Tanners Products Company, Chicago, 6 Apr. 1921, 123/1, ROH I; Rhoads transcript, 71, 105.

39 Haas to Wallin, 6 Apr. 1921.

40 Haas to Carl Immerheiser and Immerheiser to Haas, both 8 Apr. 1921, 51/3; "Agreement and License by and between BASF and Adolf Kuttroff and Röhm & Haas Company," 30 Apr. 1921, 51/3; Wallin to Haas, 17 May 1921, 123/1, ROH I.

41 Haas to Wallin, 6 Apr. 1921 ("merit"); Wallin to Haas, 8 Apr. 1921, 123/1, ROH I.

42 Wallin to Haas, 16 June 1921, and Haas to Wallin, 18 June 1921 ("laboratory"), 123/2, ROH I; Haas to Wallin, 20 Aug. 1921 ("quiet"), 123/3, ROH I.

43 Wallin to Haas, 18, 20 Aug. 1921, 123/1, ROH I.

44 Haas to Wallin, 22 Oct., 5 Nov. 1921 ("necessary"), 123/4, ROH I; Merrill A. Watson, *Economics of Cattlehide Leather Tanning* (Chicago: Rumpf, 1950), 27–28.

45 Haas to Wallin, 27 Dec. 1921, and Wallin to Haas, 30 Dec. 1921, both in 123/4; Wallin to Haas, 9 Jan. 1922, 123/5, ROH I.

46 Haas to Wallin, 15 Nov. 1922, 123/10, ROH I.

47 Wallin to Haas, 2 June 1922, and Haas to Wallin, 10 June 1922, both in 123/7, ROH I.

48 ROH, 1923 AR, [4–5]; 1924 AR, [3].

49 ROH, 1922 AR, [7].

50 Rohm and Haas Company, "Statement of Sales and Cost of Sales," 1 Jan.–31 Dec. 1926, 10/3, ROH I.

The Boss at Bridesburg

Jack J. Steelman, "Richmond-Bridesburg," in *Workshop of the World*, chap. 14; Frederick Siegle and Teresa Pyott, *Bridesburg* (Charleston, SC: Arcadia, 2004), 7–8; *The Haas Stories*, 13 (Kasperowicz), 16 ("fire").

Building Bristol

Harry Eckert, transcript of interview with M. Biddle, 23 Sept. 1982, 34/29, ROH II.

Chapter 2

1 "Philadelphia's Washington Square," *Formula* (Jan. 1959), 3.

2 ROH, 1927 AR, 8.

3 Mira Wilkins, *The Maturing of Multinational Enterprise: American Business Abroad from 1914 to 1970* (Cambridge, MA: Harvard University Press, 1974), 49–50, 68–69, 298; idem, "German Chemical Firms," 302–305.

4 ROH, 1928 AR, 7–9, and 1929 AR, 5.

5 ROH, 1921 AR, 2, and 1922 AR, [9]; Stanton C. Kelton [excerpts from Otto Haas–Otto Röhm letters on foreign operations, 1910–27], n.d. ("handing"), 8/2, ROH I.

6 ROH, 1927 AR, 8, and 1928 AR, 10.

7 Wilkins, "German Chemical Firms," 303; David A. Hounshell and John Kenly Smith, Jr., *Science and Corporate Strategy: DuPont R&D, 1902–1980* (New York: Cambridge University Press, 1988), 190–209.

8 *Ciba Builds to Serve* (New York: Ciba Company, 1953), 44; Hounshell and Smith, *Science and Corporate Strategy*, 190–209; "Report—Mr. Haas," 1 Sept. 1938, 40/4, ROH I.

9 Haas to Donald F. Murphy, Minoc, Paris, 29 Oct. 1953, 131/4, ROH I.

10 "Notables Sailing Include Gourand," *New York Times*, 22 Aug. 1923.

11 Regina Lee Blaszczyk, "True Blue: DuPont and the Color Revolution," *Chemical Heritage* 25 (Fall 2007), 20–25.

12 Resinous Products and Chemical Company, Annual Report, 1927, [hereafter cited as RPCC, 1927 AR], [1–2]. All RPCC annual reports are in 94/1, ROH I.

13 RPCC, 1927 AR, [1-2], and 1928 AR, [1–2]; Kelton, "Outline of the Resinous Products Company," 13 Oct. 1926, and idem, "Assignment of Ammann & Fonrobert Patent Application, No. 680,543, filed 13 Dec. 1923," 18 Jan. 1927, both in 96/3, ROH I; J. F. Bergin, "Agreement with Albert Company," 14 May 1948, 44/6, ROH I.

14 RPCC, 1927 AR, 4 ("recent"); 1929 AR, 4–5; and 1930 AR, 3–4.

15 RPCC, 1930 AR, 3–4; 1940 AR, 15; and 1941 AR, [14-15]. For Haas's belief in research labs see ROH, 1931 AR, 5.

16 *Th. Goldschmidt A.-G., Essen: Neun Jahrzehnte Geschichte einer Deutschen Chemischen Fabrik* (Essen, 1937), 116.

17 "Tego Gluefilm, Incorporated," 22 July 1934, 59/6, ROH I; William H. Cooke, Tego Gluefilm to Rogers & Whitaker, 28 July 1928, 58/6, ROH I; RPCC, 1933 AR, 5; agreement between Th. Goldschmidt Corp., NY, and ROH, 19 June 1934, 59/6, ROH I; correspondence between RPCC and Th. Goldschmidt Corp. in both New York and Essen, 1934, 59/6, ROH I; Rogers & Whitaker to Kelton, 11 Oct. 1934, 59/7, ROH I. On the appeal of the Goldschmidt deal see Kelton to RPCC Bookkeeping Dept., 22 June 1934 ,59/6, ROH I.

18 Haas to Kelton, 12 Nov. 1934; Kelton to Th. Goldschmidt Corp., Essen, 17 Nov. 1934, both in 59/7, ROH I; "Summary of Information of Tego Glue Film," [1934] ("revolutionize"), 59/6, ROH I.

19 RPCC, 1935 AR, 4, and 1939 AR, [2, 5].

20 RPCC, 1937 AR, 4, and 1938 AR, [7–8]; Williams Haynes, *American Chemical Industry: Histories of the Companies*, vol. 6 (New York: D. Van Nostrand, 1949), 22, 69–70.

21 RPCC, 1938 AR, [8], and 1939 AR, [5–6].

22 "The Resinous Products & Chemical Company," Aug. 1944, 96/2, ROH I.

23 RPCC, 1937 AR, 6.

24 For I.G. Farben see Werner Abelshauser et al., *Germany Industry and Global Enterprise, BASF: The History of a Company* (New York: Cambridge University Press, 2004), chap. 3; Peter Hayes, *Industry and Ideology: IG Farben in the Nazi Era*, 2d ed. (New York: Cambridge University Press, 2000); and Diarmuid Jeffrey, *Hell's Cartel: IG Farben and the Making of Hitler's War Machine* (New York: Metropolitan Books, 2008).

25 Otto Haas, handwritten notes on visits to Röhm and Haas, Darmstadt, and I.G. Farben, Aug.-Sept. 1936, 40/3, ROH I.

26 "Report—Mr. Haas," 20 Sept. 1923, 51/6, ROH I; "Report—Mr. Kirsopp," 20 Jan. 1934, 51/7, ROH I; Haas to Immerheiser, BASF, Ludwigshafen, 8 Jan. 1925, 53/3, ROH I; Haas to Patent Abteilung, BASF, 12 Jan. 1925, 53/3, ROH I; BASF to ROH, [summary of Orthochrom agreement], 1 Dec. 1934, 53/3, ROH I; Immerheiser, affidavit, May 1925, 52/1, ROH I.

27 ROH, 1929 AR, 7; "Report—Mr. Haas: General Dyestuff Corporation, New York City," 16 Mar. 1936, 53/1, ROH 1; "Syntan Royalties Paid to I.G. Farbenindustrie, 1928–1936," 52/7, ROH I. In 1928 ROH paid the I.G. $61,374 in royalties on net syntan sales of $409,165; in 1936 ROH paid $77,614 on net sales of $526,948.

28 ROH, 1932 AR, 6.

29 Haas to Walter H. Duisberg, New York, 26 July 1935, 53/1, ROH I.

30 ROH, 1929 AR, 8–10; "Review of Progress in the Sale of Echtdeckfarben during 1929" and "Review of Recent Progress in Extending the Use of Synthetic Tans," both 29 Oct. 1929, 52/3, ROH I; Thomas Blackadder, "Syntan Agreement," 28 Dec. 1937, 52/7, ROH I.

31 Drs. Holdermann and Kleber, "Report re: Second Visit of Mr. Otto Haas of Röhm and Haas Company, Philadelphia, on the 16th day of Oct. 1934 to the Patent Department in Ludwigshafen," translation, 19 Oct. 1934, 50/1, ROH I; "Translation of Protocol of October 30, 1934, covering Mr. Haas' Discussions with the I.G. Farbenindustrie," 50/1, ROH I; "Translation, Agreement Relating to Acrylic Acid and Acrylic Compounds between Röhm and Haas Company,

Philadelphia, and I.G. Farbenindustrie Aktiengesellschaft, Frankfurt a/m, October 30, 1934, and Supplementary Agreement," 50/1, ROH I; "Translation, Supplementary Agreement, May 16, 1940," 50/1, ROH I; "Report—Mr. Haas: I.G.-DuPont-Acrylic and Methacrylic Compounds," 24 Apr. 1936, 50/3, ROH I; H. T. Neher to Haas, 14 May 1936, 50/3, ROH I.

32 "Report—Mr. Haas," 9 Nov. 1934, 50/1, ROH I.

33 "Report—Mr. Haas: I.G. Farbenindustrie, AG, Ludwigshafen, Coloristische Abteilung," 9 Sept. 1936 (nos. 9–10), all in 40/2, ROH I.

34 ROH, 1936 AR, [7]; "Report—Mr. Haas: Röhm & Haas, AG, Darmstadt," 20 Aug. 1936 (no. 1a), 10 Sept. 1936 (nos. 3, 4, 5), all in 40/2, ROH I.

35 "Report—Mr. Haas: I.G. Farbenindustrie, Ludwigshafen," 30 Aug. 1938 ("new units"), 40/4, ROH I; ROH, 1941 AR, [8]; "Report—Mr. Haas: I.G. Farbenindustrie, Frankfurt," 8 Sept. 1938, 40/4, ROH I; Abelshauser et al., *Germany Industry and Global Enterprise, BASF*, 261.

36 "MAY" to I.G. Farbenindustrie AG, Coloristische Abteilung, Ludwigshafen, "Syntan Developments during the First Half of 1936," 31 July 1936, 40/2, ROH I; "Report—Mr. Haas: Exchange of Experience with I.G. on Manufacture of Mixed Polymers," 12 Sept. 1938 ("letting"), 40/4, ROH I.

37 Erwin R. Sauter to RPCC, 7 Sept., 27 Nov. 1932, 9 Jan., 22 May 1933, all in 57/16, ROH I; E. C. B. Kirsopp, "Dr. E. R. Sauter," 15 May 1933, 57/16, ROH I; "Report—Mr. Haas: Dr. E. R. Sauter," 13 Oct 1933, 57/16, ROH I; Kelton to E. R. Sauter, 19 Jan. 1934, 57/16, ROH I.

38 "Report—Mr. Haas: I.G. Farbenindustrie, Ludwigshafen, Coloristische Abteilung," 29 Aug. 1938 ("pressure"); "Report—Mr. Haas: I.G. Farbenindustrie, Frankfurt," 20 Aug. 1938; "Report—Mr. Haas: I.G. Farbenindustrie, Ludwigshafen," 19 Aug. 1938 ("admits"), all three in 40/4, ROH I.

39 Kirsopp, "Report—Mr. Kirsopp: Ciba Company, New York City," 31 Mar. 1937 ("complex"), and idem, "Urea Formaldehyde Varnishes for the Textile Trade," 17 Apr. 1934, both in 45/1, ROH I.

40 "Insecticide Report #13—C. H. Peet—February 8, 1933," 10 Feb. 1933, 42/4, ROH I; C. H. Peet to H. R. Webber, ARCO, 3 May 1934; "Atlantic Refining Co.—Lethane Agreement," 12 Oct. 1934; Kelton to H. R. Webber, ARCO, 24 Dec. 1934; Kirsopp to T. G. Delbridge, ARCO, 1 Mar. 1935; R. L. Lindabury, memorandum [on ARCO in Africa, Brazil, and Uruguay], 1 July 1936; Kirsopp to C. E. Andrews, "V.I. Improver," 17 Aug. 1936, all in 42/5, ROH I.

41 Wilkins, "German Chemical Firms," 295.

42 ROH, 1936 AR, [6], and 1939 AR, 8–9.

43 RPCC, 1940 AR, 9 ("heavy"); "Marines Have Landed—On Plywood," *Rohm and Haas Reporter* [customer magazine, hereafter cited as *Reporter*] (Dec. 1943), 6; "Urea Formaldehyde Resins Improve Shipping Containers," *Reporter* (July 1944), 11–12.

Into Leather

Turley transcript (all quotations); "Retired Chemist Remembers Early Days at Bristol," *Formula* (June 1981), 1–2.

Fashion's Wheels Turn . . . with Chemistry's Help

Regina Lee Blaszczyk, "The Color of Fashion," *Humanities: The Magazine of the National Endowment for the Humanities* (Mar.-Apr. 2008), 30–34; Hochheiser, *Rohm and Haas*, 26; ROH, 1924 AR, 7 ("Formopon"); ROH, 1926 AR, 3 ("light shades"); ROH, 1940 AR, 20 (textile chemicals as 31 percent of gross profits); Rohm and Haas, *Chemicals for Industry* (Philadelphia: Rohm and Haas, 1945), 52–77.

Chapter 3

1 Robert W. Rydell et al., *Fair America: World's Fairs in the United States* (Washington, DC: Smithsonian Institution Press, 2000), 90–96.

2 Rohm and Haas spent $47,000 on the fair, nearly half the net profit on plastics for 1939; ROH, 1939 AR, [13].

3 Stephen Fenichell, *Plastic: The Making of a Synthetic Century* (New York: Harper Business, 1996), 147; Helen A. Harrison et al., *Dawn of a New Day: The New York World's Fair, 1939/40* (Flushing, NY: Queens Museum, 1980), 52, 54.

4 *The General Motors Exhibit Building*, pamphlet, 1939, author's collection; *Official Guide Book of the New York World's Fair*, 1st ed. (New York: Exposition, 1939), 209; J. Breedis, "Monthly Report #12," Bristol, 11 Jan. 1939, 37/2, ROH I. The Plexiglas "X-Ray Car" was such a hit that Fisher created a new model for the 1940 fair. For Don Frederick's response to the fair see Frederick transcript, 19, 27.

5 ROH, 1932 AR, 8.

6 Jeffrey L. Meikle, *American Plastic: A Cultural History* (New Brunswick, NJ: Rutgers University Press, 1995), 82–90; Ralph L. Woods, "Great Promise in Chemistry's Latest Progeny," *Magazine of Wall Street* 57 (7 Dec. 1935), 201–202, 231–232.

7 Richard S. Tedlow, *New and Improved: The Story of Mass Marketing in America* (Boston: HBS Press, 1996), chap. 3.

8 Joseph Rossman, "The Manufacture of Laminated Safety Glass: Part I," *Glass Industry* 10 (Jan. 1929), 1–6; E. Ward Tillotson, "Automobile Glasses," *Glass Industry* 8 (Dec. 1927), 282–284; F. G. Schwalbe, "How Ford Makes Its Glass: III," *Glass Industry* 19 (Feb. 1938), 35–60.

9 *100 Years of Advancement: Röhm GmbH from 1907 to 2007* (Darmstadt, Germany: Röhm GmbH, 2007), 27–29.

10 "American Window Glass Company," typed excerpts from Otto Haas letters to Otto Röhm, 1928–30, n.d., 26 Mar., 30 Apr., 18, 19 June 1928, 41/5, ROH I.

11 "American Window Glass Company," 26 Dec. 1928.

12 *100 Years of Advancement,* 29–31.

13 *100 Years of Advancement,* 32.

14 "Report—Mr. Haas," 12 Sept. 1931, 45/7, ROH I ("financial difficulties"); Kelton, "Note for Bookkeeping," 2 Feb. 1932, 45/7; Kelton, "Memo for Mr. Haas," 19 Sept. 1934, 45/9, ROH I. Haas described the laminated glass development from "our Darmstadt friends" in ROH, 1931 AR, 6.

15 Kelton, "Note for Mr. W. C. Becker," 7 Mar. 1932; Haas to Röhm, 25 Apr. 1932 ("business"); Haas, "Plexigum—Payments to Darmstadt," 12 May and 4 June 1932, all in 45/7, ROH I. Röhm first thought about selling out to the "little I.G." in 1920–21; see Kelton [excerpts from Haas-Röhm letters on foreign operations, 1910–27], n.d., 8/2, ROH I.

16 ROH, 1933 AR, 11–12; 1934 AR, 4; 1935 AR, [3–4]; and 1936 AR, 7–9; "Report—Mr. Haas," 29 Sept. 1933, 41/5, and "Report—Mr. Haas," 23 Sept. 1933, 41/9, ROH I; *Plexite: The Safer Safety Glass* (Pittsburgh: American Window Glass Company, 1936–41], Trade Catalog Collection, Smithsonian Institution Libraries, Washington, DC.

17 *100 Years of Advancement,* 35–42; Donald S. Frederick, "Miscellaneous Items Concerning Plexiglas Manufacture," 2 Mar. 1936, 2, 36/5, ROH I.

18 Frederick transcript, 18; O. P. Echols, "Transparent Panels in Cockpit Enclosures," 2 Sept. 1936, box 3436, RG 342, National Archives and Records Administration, Archives II, College Park, MD [hereafter cited as NA].

19 Donald S. Frederick, "Manufacture of Plexiglas Sheets," 6 Feb. 1936; Frederick to Haas, 11 Feb. 1936; Frederick, "Report of Discussions with Drs. Röhm, Bauer, Weisert, and Kautter," 11 Feb. 1936; Frederick, "Preparation of Monomeric Methyl Methacrylate," 11 Feb. 1936; Frederick, "Discussions in Darmstadt," 2 Mar. 1936; Frederick, "Miscellaneous Items Concerning Plexiglas Manufacture," 2 Mar. 1936; E. L. Helwig, "Improvement and Development Report for the Month of January, 1936," 3 Feb. 1936, all in 36/5, ROH I.

20 S. L. Blackman, "A Long and Short View of Aviation's Future," *Magazine of Wall Street* 57 (29 Feb. 1936), 566.

21 Charles M. Turner, "Aviation Begins to Earn Money," *Magazine of Wall Street* 58 (10 Oct. 1936), 760–762, 798; Echols to Edgar S. Gorrell, Air Transport Association of America, Chicago, 19 Sept. 1936, box 3856, RG 342, NA.

22 Gorrell to Maj. Gen. O. Westover, War Dept., Washington, DC, 29 Aug. 1936, box 3856, RG 342, NA.

23 Col. C. G. Hall, acting chief of the Air Corps, to Gorrell, 3 Sept. 1936, box 3856, RG 342, NA.

24 Bjorn Andersen, Celluloid Corporation, to Maj. F. O. Carrroll, Wright Field, 15 Feb. 1937, box 8014, RG 342, NA; Maj. John M. Davies, Hamilton Field, to Materiel Division Chief, Wright Field, 16 Mar. 1937, box 3856, RG 342, NA.

25 ROH, 1936 AR, 9, and 1937 AR, [9] ("handled"). Haas marveled at the "swift" pace of the plastics industry in ROH, 1939 AR, 11.

26 ROH 1940 AR [6]; Williams Haynes, *American Chemical Industry*, vol. 6 (New York: D. Van Nostrand, 1947), 358; "Bristol Plant: Changes and Improvement during 1936," "Bristol Plant: Changes and Improvements during 1937," and "Annual Report of the Bristol Plant," 12 Feb. 1940, all 135/8, ROH I.

27 Frederick, "Report 37-48: Call on the American Optical Company, Southbridge, Mass., Mar. 25, 1937"; idem, "Report 37-45: Call at the General Electric Company, Schenectady, NY, Mar. 23, 1937"; idem, "Report 37-48: Call at the General Electric Company, Pittsfield, Mass., 24 Mar. 1937," all in 36/5, ROH I.

28 For a Plexiglas order from Douglas Aircraft see Frederick to E. L. Helwig, 12 June 1937, 36/5 ROH I. On high-altitude planes see Frederick to J. B. Johnson, Wright Field, 11 and 30 Apr. 1937, box 3856, RG 342, NA; W. F. Bartoe, "Monthly Report #3: Physical Laboratory Report for April 1937," 4 May 1937; 37/2, ROH I. Research on high-altitude applications continued during the war; ROH, AR 1941, [12].

29 Frederick to Wright Field, 9 May 1941, box 3856, RG 342, NA; J. E. Schaefer, Stearman Aircraft, to Paul H. Kemmer, Wright Field, 13 Aug. 1938, box 3856, RG 342, NA.

30 Frederick to Maj. Ray C. Dunn, Middletown Air Depot, 1 July 1938

("attempting"); Frederick to Johnson, 12 July 1938 ("recently"), box 3856, RG 342, NA; Schaefer to Kemmer, 13 Aug. 1938, box 3856, RG 342, NA.

31 ROH, 1940 AR, [14]; 1942 AR, [6]; 1943 AR, 23; and 1944 AR, 15.

32 Frederick transcript, 16, 164; ROH, 1937 AR, [9]; 1940 AR, [14, 20]; and 1941 AR, 5.

33 ROH, 1943 AR, 14, 20 ("normal times").

34 ROH, 1941 AR, [7].

35 ROH, 1942 AR, 11 ("South Gate"). Aircraft companies sent dozens of telegrams thanking Rohm and Haas for help with Plexiglas; see, for example, Glenn L. Martin Co., Baltimore, to ROH, Bristol, 19 July 1942 ("many thanks"), box 1, Photograph Accession, Rohm and Haas Archives, Philadelphia, PA [hereafter cited as ROH P].

36 ROH, 1946 AR, 49; C. B. Wooster, "Report No. 43-5: Investigation of Plant Operating Difficulties at Knoxville," 3 Nov. 1943, 37/2, ROH I; "New Production Facilities for Acrylics," *Reporter* (Feb. 1943), 5; "Unusual Plant Doubled Plexiglas Output," *Reporter* (Sept. 1944), 1–3.

37 "Application for 'Excellence Award,' " July 1942, box 1, ROH P; ROH, 1945 AR, 65.

38 ROH, 1944 AR, 16, and 1945 AR, 17.

39 ROH, 1945 AR, 8, 13, 17.

40 Frederick transcript, 18.

41 ROH, 1938 AR, 13.

42 Otto Haas, "Report—Mr. Haas: Darmstadt Visit, Aug. 9–19," 19 Aug. 1939, 40/6, ROH I; "Anti-Trust Indictments Involving Röhm & Haas Company," 29 Sept. 1943, 16/2, Accession II, Rohm and Haas Archives, Philadelphia, PA [hereafter cited as ROH II]; F. J. Rarig to J. T. Subak, 29 July 1981 (draft letter to U.S. Attorney General William French Smith); Rarig to Subak, 16 Oct. 1981 (draft letter to William F. Baxter, Asst. Attorney General); James Fay Hall, Jr., to Subak, 23 Oct. 1981; Subak to Jeffrey I. Zuckerman, U.S. Justice Dept., 4 Dec. 1981, all in 13/5, ROH II; J. S. Stroebel, "Vacation of R&H Consent Decree," 21 June 1982, 13/6, ROH II. The consent decree was vacated in Dec. 1983; see ROH, 1983 AR, 2.

43 ROH, 1940 AR, [14].

The Machine Age

On Rohde see boxes A and F, Gilbert Rohde Collection, Cooper-Hewitt National Design Museum, New York, NY; Carl Greenleaf Beede, "Furnishings in a New Material," *Christian Science Monitor*, 31 Oct. 1939; Phyllis Ross, "Merchandising the Modern: Rohde at Herman Miller," *Journal of Design History* 17 (2004), 359–376; idem, "A Bridge to Postwar American Design: Gilbert Rohde and the 1937 Paris Exposition," in Donald S. Albrecht, ed., *Paris New York: Design Fashion Culture, 1925–1940* (New York: Monacelli Press, 2008), 198–215; and idem, *Gilbert Rohde: Modern Design for Modern Living* (New Haven: Yale University Press, 2009), chap. 8. On the contest see folder CE II.1.47.5: Competition for Sculpture in Plexiglas, MoMA Archives, New York, NY. For the quotation on Calder see press release, May 1939, folder CE II.1.47.5, MoMA Archives. For the Kohn quotation see his letter to Frederick, 1 Mar. 1939, box 69, 1939 New York World's Fair Collection, New York Public Library, New York, NY, courtesy of Phyllis Ross.

War on Insects

Hochheiser, *Rohm and Haas*, 33–35; D. F. Murphy, "Testing Insect Sprays," *Reporter* (Feb. 1943), 3–4; C. J. Dumas, "Synthetic Insecticides," *Reporter* (Mar. 1943), 1–5; "Insect Killer for Victory Gardens," *Reporter* (July 1944), 8–9; "New Product Aids War on Pests," *Reporter* (Mar. 1944), 8–10; "Flies Drop Like Flies in Testing Laboratory," *Formula* (Dec. 1947), 3 (quotations); "Perthane . . . Bright Future for New Insecticide," *Formula* (Mar. 1946), 3; "Dithane Gives Growers Crop Insurance," *Formula* (May 1945), 3; ROH, 1952 AR, 14 (Peet-Grady).

Chapter 4

1 On Murphy's importance see J. Lawrence Wilson, transcript of oral history with James G. Traynham, 14, 30 Aug. 1999 [hereafter cited as Wilson transcript (1999)], Oral History Collection, Chemical Heritage Foundation, Philadelphia, PA [hereafter cited as CHF].

2 F. Otto Haas, transcript of oral history with Sheldon Hochheiser, 47, 13 Apr. 1984, ROH.

3 Alan Barton, interview with author, Philadelphia, 22 Sept. 2008.

4 ROH, 1930 AR, 11, and 1938 AR, 7–8 ("Oropon").

5 ROH, 1936 AR, 3.

6 ROH, 1938 AR, 7–8; "Rohm and Haas Products Reach Many Foreign Markets," *Formula* 1 (Mar. 1946), 3–10.

7 ROH, 1936 AR, 3.

8 RPCC, 1945 AR, 11.

9 ROH, 1946 AR, 38 ("good demand").

10 Hochheiser, *Rohm and Haas*, 142; ROH, 1946 AR, 1, and 1947 AR, 31.

11 Hochheiser, *Rohm and Haas*, 33–35, 109. For his agricultural work see Donald F. Murphy, "Oils for Use with Lethane Spray," 8 June 1936; "List of Insects against Which Lethane Is Effective," [8 June 1936]; and "Data Concerning the Safety of Lethane 384," 9 June 1936, all 46/3, ROH I. On the Bristol farm see *Lethane 384* (Philadelphia: Rohm and Haas, ca. 1936), 46/3, ROH I. Also ROH, 1940 AR, [7]; 1942 AR, 16; and 1944 AR, 30; ROH, Minutes of the Board of Directors Meeting [hereafter cited as ROH, BM], 17 Apr. 1951, 4–5 ("exceptional"); ROH, BM, 22 Jan. 1957, 19–20. Unless indicated otherwise, board minutes are in ROH I.

12 Murphy to Otto Haas, "Pechiney," 2 Apr. 1953 ("Mr. Moundlic"), 131/4, ROH I; [Dr. Ely M. Swisher], *Reporter* (Sept. 1944), cover; Ralph Connor, "Research: A Cornerstone of Our Company," *Formula* (May 1949), 3, 8; "Italy's Valley of Fruit," *Reporter* (Jan.–Feb. 1961), 8–11.

13 ROH, BM, 24 July 1951, 20–24; 23 Oct. 1951, 19–21; 22 Jan. 1952, 10, 16–17; and 27 July 1965, 3; Murphy, "Shell Petroleum Company," 23 Nov. 1951 ("anti-trust"), 37/4, ROH I.

14 ROH, BM, 22 Jan. 1952, 16–18 (17, "termination"), and 28 Oct. 1952, 19–20; Murphy and Frederick W. Tetzlaff, "Dithane Results—Switzerland—1951," 6–7 Dec. 1951, 37/4, ROH I; Murphy and Tetzlaff, "Conferences with French Government Officials," 3–4 Dec. 1951 (Pavot, Vesin), 37/4, ROH I.

15 Murphy, "American Embassy Discussions, Paris," 28 Nov. 1951, 37/4, ROH I; Murphy, "Professor Raucourt, Director, National Phytopharmacie Service, Versailles," [Nov. 1951] ("visa"), 37/4, ROH I.

16 Hochheiser, *Rohm and Haas*, 109; Murphy and Tetzlaff, "Conferences with French Government Officials"; "Ministry of Commerce & Energy and Ministry of Agriculture," 19 Dec. 1951 ("operations"); "Manufacturing Arrangements for Dithane in France," 29–30 Nov. 1951; "Manufacturing Arrangements for Dithane in France," 5–6 Dec. 1951 ("aggressive"); "Vernou Plant—La Quinoléine Co.," 17 Dec. 1951; and "Dithane Production in France," 18–20 Dec. 1951, 37/4, ROH I.

17 Murphy to Jean Pomot, 3 Jan. 1952, 8/7, ROH I; Bob Reitinger, transcript of oral history with Sheldon Hochheiser, 11 July 1983, 13–15, ROH.

18 Tetzlaff to Duncan Merriwether, 22 Feb. 1952; Tetzlaff to Haas, 29 Feb. 1952, 8/8, both ROH I. On the importance of being in close proximity to the customers see Tetzlaff to Haas, 30 May 1952, 131/3, ROH I.

19 ROH, BM, 15 Apr. 1952, 11, and 21 Apr. 1953, 16–17; Murphy, "Program for European Activities—1952," 25 Jan. 1952 ("outlet"), 8/7, ROH I.

20 ROH, 1952 AR, 14–15; ROH, BM, 22 July 1952, 18; Murphy and Tetzlaff, "Dithane Distribution Program for France," 26–30 Nov. 1951, 37/4, ROH I; "Lambert-Rivière," 18 Dec. 1951, 37/4, ROH I; "Progil,"18 Dec. 1951, 37/4, ROH I; and "Pest Control Ltd.," 18–19 Dec. 1951, 37/4, ROH I; Haas to Murphy, Paris, 18 Mar. 1952, 8/8, ROH I; Murphy and Tetzlaff to Haas, 12 April 1952, 131/3, ROH I; Vincent L. Gregory, Jr., transcript of oral history with James J. Bohning, 12, 14 Feb. 1995, CHF [hereafter cited as Gregory transcript (Bohning, 1995)].

21 ROH, 1952 AR, 16; Vincent L. Gregory, Jr., transcript of oral history with Sheldon Hochheiser, 12, 23 Mar. 1984, ROH [hereafter cited as Gregory transcript (Hochheiser, 1984)].

22 Elisabeth Timper files, Feb.–Oct. 1954, 131/5, ROH I.

23 ROH, 1952 AR, 15; Tetzlaff, "Simonnot, Rinuy, Blundell & Pont, Paris," 26 Mar. 1953 ("French market"), 8/8, ROH I.

24 Tetzlaff, "Pechiney French Distribution," 25 Mar. 1953 ("sold"), 8/8; and "Dithane in France," 23 Oct. 19 1953, 37/5, ROH I.

25 A. Vesco to Minoc, with "Note on the Commercial Organization of Sodec," 12 Nov. 1953; M. H. J. Villeneuve, "Progil," 8 Sept. 1953, both in 8/8, ROH I. Also Tetzlaff, "Dithane in France," 23 Oct. 1953; Tetzlaff, "French Distribution," 15 Nov. 1953 ("expectations"); and Murphy and Tetzlaff, "Dithane in France," 10 Nov. 1953, all 37/5, ROH I.

26 Tetzlaff and Murphy, 16 Mar. 1954 ("strange") and 15 Mar. 1954 ("nationals"), 131/5, ROH I.

27 "Field Testing on Company Farm," *Formula* (May 1957), 10; Ralph Connor to Murphy, "Dr. R. Maag, AG—Switzerland—Formulation Research," 3 Jan. 1957, 37/8, ROH I; ROH, AR 1952, 13–14.

28 Basil A. Vassiliou, personal communication to author, 26 Sept. 2008; Murphy, "Dr. R. Maag & Co.," 9 Nov. 1955, 37/5; "Dr. R. Maag Company," 16 Nov. 1955, 37/6; "Dr. R. Maag, AG," 18 Feb. 1956, 37/7; "Dr. R. Maag Company," 6 June 1957, 37/10, all ROH I.

29 Murphy, "Dr. R. Maag & Co.—Switzerland—Relations," 9 Nov. 1953, 37/5, ROH I.

30 ROH, BM, 22 July 1958, 8–9, 21/4, ROH I; Gregory transcript (Bohning, 1995), 18; Vassiliou, personal communication to author.

31 Basil A. Vassiliou, transcript of oral history with Regina Lee Blaszczyk, Dec. 2006, ROH [hereafter cited as Vassiliou transcript]; "Dr. G. Assalini," 26 June 1957, 37/10, ROH I.

32 ROH, BM, 28 Jan. 1958, 11.

33 Ralph Connor to Murphy, "C. H. Boehringer Sohn," 29 Aug. 1958, 8/9, ROH I.

34 "Golden Anniversary, 1959," *Formula* (Aug. 1959); ROH, 1952 AR, 16; ROH, BM, 22 July 1952, 6–7; 28 Oct. 1952, 7; 28 July 1953, 16–17 ("dumping laws"); 26 Jan. 1954, 4; and 21 Apr. 1964, 11.

35 ROH, BM, 23 Oct. 1956, 19, and 26 Jan. 1960, 9.

36 ROH, BM, 28 Oct. 1952, 22 ("valuable"); 26 Jan. 1954, 18; 19 Apr. 1955, 17; 24 July 1956, X–3; and 26 Jan. 1960, 12–13.

37 ROH, BM, 22 July 1952, 18–19; 28 Oct. 1952, 24–25; 27 Jan. 1953, 13; 28 July 1953, 16; 26 Jan. 1954, 17–18 (18, "Finance Ministers"); 26 Oct. 1954, 10; 19 Apr. 1955, 11 ("key equipment"); 24 July 1956, X–3; 28 Oct. 1958, 18; 27 Jan. 1959, 26; 26 Jan. 1960, 11–12 (12, "despite"); 18 Apr. 1961, 12–13; and 22 Jan. 1963, 18.

38 ROH, BM, 22 July 1952, 19; 24 July 1956, X3–X4; 25 Oct. 1955, 21; 28 Jan. 1958, 21; and 15 Apr. 1958, 17.

39 Peter Binzen, "The Journey of Rohm & Haas' American in Paris," *Philadelphia Inquirer*, 27 June 1988; "The End of an Era: Gregory to Retire," *Formula* (June 1988), 1, 3, 4; James W. Michaels, "Kindred Spirits," *Forbes* (18 Oct. 1999), 138–139.

40 Gregory transcript (Hochheiser, 1984), 4–5.

41 ROH, BM, 26 Oct. 1954, 10; Vincent L. Gregory, Jr., transcript of interview with Steve Ozer, 31 May 1988, 8–9 ("going around") [hereafter cited as Gregory transcript (Ozer, 1988)], ROH; Gregory transcript (Hochheiser, 1984), 7–8.

42 Gregory transcript (Ozer, 1988), 7–8.

43 For Murphy's influence on Gregory see "The End of an Era," 3; for "push," see Gregory transcript (Ozer, 1988), 8.

44 "The End of an Era," 3; Gregory transcript (Hochheiser, 1984), 8.

45 ROH, BM, 17 Apr. 1951, 4–5.

46 Gregory transcript (Hochheiser, 1984), 11.

Passing the Baton

Hochheiser, *Rohm and Haas*, chaps. 9–10; "Otto Haas," *Formula* (Jan. 1960), n.p.; Gregory transcript (Bohning, 1995), 11.

Give Us Ca or Mg

"New Ion Exchange Resins Announced to Technical Press," *Formula* (July 1948), 3; "Laboratory 36 Explores Many Fields for Use of New Ion Exchange Resins," *Formula* (Mar. 1949), 8; Robert Kunin, "Six Decades of Ion Exchange Technology at Rohm and Haas," *Chemical Heritage Magazine* 17 (Summer 1999), 8–9, 36–41; ROH, BM, 25 July 1950 (Otto Haas quote); Vassiliou transcript, 2–4.

Chapter 5

1 "Explorer III—Eye-Opener in Detroit," *Reporter* (Mar.-Apr. 1965), 16–19; *Explorer III* (Philadelphia: Rohm and Haas, 1965), ROH.

2 On Schmidt's background see William M. Schmidt Papers, acc. 1672, Benson Ford Research Center, Henry Ford Museum, Dearborn, MI; "Automotive Designs in Plexiglas," *Reporter* (Jan.-Feb. 1963), 16–17; Dwight L. Pickard, "It's No Plane, It's No Bird, It's Supersalescar!" *Philadelphia Sunday Bulletin Magazine*, 10 Aug. 1969, 8–11 (11, "taillights"), 79/3, ROH I.

3 *Explorer IV* (Philadelphia: Rohm and Haas, 1967), ROH; "Happy Birthday, Plexiglas," *Reporter* (Winter 1988), 31; "Plexiglas on the Highways," *Reporter* (Apr. 1949), 10–15; Rohm and Haas, *Chemicals for Industry, 1909–1959* (Chicago: Lakeside Press, 1959), 8–20; "Your Safety Is Their Business," *Reporter* (Jan.-Feb. 1965), 4–7.

4 Blaszczyk, *American Consumer Society*, pt. III.

5 Ralph Connor, "Research," *Formula* (Jan. 1960), 6.

6 ROH, 1947 AR, 11; ROH, BM, 21 Apr. 1959, 14; Lloyd W. Covert, "1948 in Review," *Formula* (Jan. 1949), 3–4; Wilson transcript (1999), 13; "Instruction for Army and Navy," *Reporter* (Dec. 1943), 8–10; ROH, 1945 AR, 17. Charles Lennig and Company was dissolved and its assets and liabilities transferred to ROH on 31 Dec. 1947; RPCC was merged into ROH on 14 Sept. 1948.

7 "Plexiglas Promotion in Consumer Field," *Reporter* (Sept. 1944), 12.

8 ROH, 1946 AR, 12, 40 ("many"), 49, and 1947 AR, 14, 16; advertisement, "What Can You Do with Plexiglas," *Reporter* (Dec. 1944), verso front cover; "Packaged in Plexiglas," *Reporter* (Mar. 1946), 1–3.

9 On Frederick's reorganization of the Plastics Department see ROH, 1946 AR, 41; ROH, BM, 28 July 1953, 4 ("use").

10 ROH, 1945 AR, 16 ("wide"); Stanton C. Kelton, Jr., transcript of oral history with Sheldon Hochheiser, 24 June 1983, 23 ("colored"), ROH; Kelton, "Acrylic Color Techniques," typescript of article for *Modern Plastics*, Aug. 1954, and "Colorants for Acrylic Plastics," typescript of paper presented at the Perkin Centennial, 11 Sept. 1956, both in 9/15, ROH II; ROH, 1947 AR, 37–38.

11 Kelton transcript, 23–25.

12 Ibid., 31 ("machines" quotation and the demand for red taillights); ROH, 1952 AR, 6–7.

13 ROH, 1945 AR, 61 ("range"); Mary Roche, "Home Highlights for the Month," *New York Times*, 16 Sept. 1945; "Plexiglas in the Postwar Home," *Reporter* (Oct. 1945), 1–2, 6–7, 12; ROH, 1946 AR, 15; Frederick transcript, 114.

14 F. W. Tetzlaff, "Acrylic Plastics in the Lighting Field," *Chicago Electrical News* (Sept. 1948), offprint, 92/3, ROH I; "Plexiglas Helps Bring Stores Up-to-Date," *Reporter* (July 1948), 8–9; "Plexiglas as a Building Material," *Reporter* (Feb. 1949), 1–3, 24–25; "Store Fronts Go Modern," *Reporter* (June 1949), 10–11; "A Sign of Success," *Reporter* (May-June 1963), 24–27.

15 "Why You Should See 'The Sign of Plexiglas,' " *Reporter* (Jan.-Feb. 1960), 14–15.

16 ROH, 1952 AR, 4–7, 22, [35–39] (6, "weather").

17 "Acrylic Signs Threaten Neon," *Modern Plastics* (Jan. 1950), offprint, 92/3, ROH I; "Daytime Gleam—Nighttime Glow," *Reporter* (Feb.-Mar. 1950), 11–15; Rohm and Haas, *Chemicals for Industry, 1909–1959,* 71–73. Customer Charles Traipe described the role of signs in branding in "A Sign of Success," *Reporter* (May-June 1963), 27.

18 Also see "Plexiglas Users Win Competition Awards," *Reporter* (Sept. 1948), 8–9; "Signs Sell Service," *Modern Plastics* (Jan. 1950), offprint, 92/3, ROH I; "Big Trade-Mark," *Reporter* (Nov.-Dec. 1953), 12–13, 28.

19 "Plexiglas for Signs," *Reporter* (Mar. 1948), 8–9; "Signs of Progress," *Reporter* (Nov.-Dec. 1954), 14–18; "Signs Sell Service," *Reporter* (May-June 1961), 16; "Big Change in Signs," *Reporter* (July-Aug. 1961), 4–5; "High Sign for Motorists," *Reporter* (Jan.-Feb. 1967), 29–31; "Geared to a Modern America," *Reporter* (July-Aug. 1965), 10–13; "A Sign on the Go for Atlantic," *Reporter* (May-June 1965), 29–31.

20 "Lots of Showmanship," *Reporter* (June-July 1950), 10–13; "A Sign of Ford Country," *Reporter* (Mar.-Apr. 1967), 4–8; "Symbols Sculptured in Plexiglas," *Reporter* (May-June 1968), 4–8; "Plexiglas for 'The Mark of Excellence,' " advertising copy, Dec. 1968, 81/3, ROH I; "A New Sign of Excellence," *Reporter* (July-Aug. 1969), 8–11.

21 "Spreading the Story of Plexiglas," *Formula* (Sept. 1958), 3; "What's Back of This Sign," ad copy for *National Petroleum News*, 1 Apr. 1968, 81/2; "New Handbook for the Seventies" and "The Rohm and Haas Sign Seminar," ad copy, both 27 Oct. 1969, 81/5, ROH I.

22 "Trademarks in Plexiglas," *Reporter* (July-Aug. 1964), 16–20.

23 "A Vital Ingredient of Success," *Reporter* (Jan.-Feb. 1965), 22–25.

24 Robert J. Whitesell, transcript of oral history with Sheldon Hochheiser, 19 Sept. 1983, 33 ("by 1960"), ROH [hereafter cited as Whitesell transcript]; ROH, BM, 16 Apr. 1963, 35–38; Fred W. Shaffer, transcript of oral history with Regina Lee Blaszczyk, 1 Dec. 2006, 13, ROH [hereafter cited as Shaffer transcript]; Ralph Connor, transcript of oral history with Sheldon Hochheiser, 29 Sept. 1983, 32–33 ("to do it"), 66–72, ROH [hereafter cited as Connor transcript].

25 ROH, BM, 23 Jan. 1962, 18 ("acquisitions"); Connor transcript, 32–33 (quotations); Whitesell transcript, 33; press release, "Rohm and Haas Company as a Specialty Fiber Producer," 2 May 1971, 79/8, ROH I.

26 ROH, BM, 23 Jan. 1962, 19; Louis Klein, transcript of oral history with Sheldon Hochheiser, 18, 21, 13 July 1983, ROH [hereafter cited as Klein transcript]; Shaffer transcript, 12.

27 John C. Haas, transcript of oral history with Sheldon Hochheiser, 8 Mar. 1984, 29 ("copolymers"), ROH [hereafter cited as John C. Haas transcript]; ROH, BM, 26 Oct. 1970, 10.

28 ROH, BM, 24 July 1962, 12–13; 23 Oct. 1962, 14–15; 22 Jan. 1963, 20–21; 16 Apr. 1963, 15–16; 23 July 1963, 9–10; 22 Oct. 1963, 7; 28 Jan. 1964, 10; 21 Apr. 1964, 12; 28 July 1964, 10; and 26 Jan. 1965, 16–17 ("developed"); "Rhee Industries Joins Rohm & Haas," *Reporter* (Mar.-Apr. 1962), 1; "Sauquoit Silk Acquired by Rohm & Haas," *Reporter* (Mar.-Apr. 1963), 1.

29 ROH, BM, 27 Oct. 1964, 14–15; 20 Apr. 1965, 20–22, 28; 27 July 1965, 2–4, 7, 17; 26 Oct. 1965, 25, 34–35; 25 Jan. 1966, 17–18; 19 Apr. 1966, 19–21; 26 July 1966, 3–5, 28–31 (Tetzlaff's quotations); and 24 Jan. 1967, 30.

30 "Unusual Laboratory Aids in Plexiglas Development," *Reporter* (Mar. 1945), 1–2.

31 Frederick transcript, 142–143 (Frederick quotations); "New Look for Refreshment," *Reporter* (July-Aug. 1958), 12–15; press release, 16 Dec. 1959, box 17, Series IIA, Raymond Loewy Collection, HML; "Promoting More Sales," *Reporter* (Mar.-Apr. 1961), 21–24; "Plexiglas—A Practical New Architectural Material," *Reporter* (Jan. 1948), 8–9.

32 Frederick W. Tetzlaff and Robert R. Rorke, "Acrylic Plastics in Architecture," *Progressive Architecture* (July 1949), 75–78, offprint, 92/3, ROH I; "Lighting That's Easy to Live With," *Reporter* (July-Aug. 1965), 4–7; "A New Development in Street Lighting," *Reporter* (Jan.-Feb. 1953), 1–5; "Lighting the Highways for Safety," *Reporter* (Jan.-Feb. 1958), 12–16; Plastics Department, 1968 Annual Report, 4/4, ROH I; "Leader in Creative Lighting," *Reporter* (Nov.-Dec. 1963), 19–23 ("designer's dream"); "Building with Plexiglas," *Reporter* (Jan.-Feb. 1964), 16–20 ("belong").

33 "Focus on the Fair," *Reporter* (May-June 1962), 16–19; "Plexiglas at the Fair," *Reporter* (Sept.-Oct. 1964), 10–19, back cover; "Pavilion of Plexiglas," *Reporter* (Mar.-Apr. 1967), 24–27; "An Exposition in Plexiglas," *Reporter* (Nov.-Dec. 1967), 24–28.

34 ROH, BM, 18 Apr. 1961, 18, and 23 Oct. 1962, 20–21; "Private Building Respects Public Site," *Architectural Record* (Jan. 1966), offprint; "Rohm & Haas," *Contract* (Apr. 1966), 80–89, both in 148/4, ROH I; "New Art Forms in Plexiglas," *Reporter* (Mar.-Apr. 1966), 2–3, 6–11; *Elegance on the Mall* (Philadelphia: Rohm and Haas, 2008).

35 Plastics Department, 1968 and 1970 Annual Reports, 4/4, ROH I; ROH, BM, 24 Jan. 1967, 9, and 25 Oct. 1966, 12–13.

36 ROH, BM, 28 July 1969, 27, and 27 Oct. 1969, 31–32; press release, "First New Fiber in a Decade," 22 Oct. 1969, 79/8, ROH I; "Anim/8," *Reporter* (Jan.-Feb. 1970), 4–7. The promotion cost $350,000.

37 Delores Cree, Rogers, Cowan and Brenner, New York, to Mike Storti, 19 Sept. 1969, 79/8, ROH I; Laura H. Stevenson, "Bob & Deanna Littell Exile Themselves into a Joyous Life," *People* 4 (28 July 1975).

38 "First New Fiber in a Decade."

39 Press release, "Rohm and Haas Company as a Specialty Fiber Producer," 2 May 1971; press release, "New Yarn That Eliminates Cling in Double Knit Fabrics Available from Rohm and Haas," 2 May 1971; flyers for "Formelle," "X-Static," and "Anim/8," [1971], all in 78/8, ROH I: "Panty Hose with Plus Performance," *Reporter* (Winter 1971), 12–15; ROH, BM, 26 Oct. 1970, 21; publicity photo for Patternskins, [1971], 78/8, ROH I.

40 ROH, BM, 23 Jan. 1968, 45–46; 22 Oct. 1968, 58–59; 28 Jan. 1969, 45; 28 Apr. 1969, 37 ("Enka"); 26 Jan. 1970, 41–42; 27 Apr. 1970, 21–22 (21, "giants"); 27 July 1970, 37–40; and 26 Oct. 1970, 24, 27–28.

41 Shaffer transcript, 14; J. Lawrence Wilson, transcript of oral history with Regina Lee Blaszczyk and Arnold Thackray, 16 Aug. 2006, 28, ROH [hereafter cited as Wilson transcript (2006)]; Hochheiser, *Rohm and Haas*, 161–164.

42 Donald L. Felley, transcript of oral history with Sheldon Hochheiser, 24 Feb. 1984, 34–36 (Felley quotations); Fred W. Shaffer to Vincent L. Gregory, "Discussion of Our 1977–1979 Business Plan," 4 May 1977, and "Minutes of Management Committee Meetings," 27–28 June 1977, both in Management Committee Meetings Binder [hereafter cited as MCM], ROH.

43 "Underwriters Test and Approve Plexiglas," *Reporter* (Nov. 1948), 14–15; ROH, BM, 24 Oct. 1950, 22, and 26 Oct. 1954, 27; "Peaceful Code Existence," draft copies for *Plexiglas Sales Letter (Sign Supply Co.)* and *Plexiglas Sales Letter (Distributor)*, both 18 Aug. 1967, 80/12, ROH I; Frederick J. Rarig, "Taking Smoke's Measure," *Reporter* (Sept.-Oct. 1970), 9–11; "Unusual Laboratory Aids in Plexiglas Development"; "Color in Plastics," *Reporter* (May-June 1960), 8–11; "A Laboratory for Users of Plexiglas," *Reporter* (Nov.-Dec. 1964), 20–23.

44 John C. Haas transcript, 31.

45 Ibid., 29 ("kibosh"); ROH, BM, 25 Oct. 1966, 24.

Diversification Decade

F. Otto Haas, oral history by Sheldon Hochheiser, 13 Apr. 1984, 46–47 (quotations); Gregory transcript (Bohning, 1995), 21 (quotation); Hochheiser, *Rohm and Haas*, chaps. 11, 13.

South of the Border

Garaventi transcript, iii, 6, 15–20 (16, Hank, Lois, Colombia; 19–20, "lessons").

Chapter 6

1 Vassiliou transcript, 16.

2 Ellington Beavers, transcript of oral history with Sheldon Hochheiser, 12 Aug. 1983, 66, ROH.

3 ROH, 1946 AR, 44–56, and 1947 AR, 40–41 ("manufacturing"); Lloyd W. Covert, "1948 in Review," *Formula* (Jan. 1949), 3–4 ("hydrogen cyanide").

4 Connor transcript, 16–17; ROH, "Acrylic Emulsion Technology by Rohm and Haas," nomination to the American Chemical Society for landmark status,

Oct. 2007, 3; ROH, BM, 26 July 1949, 11 ("exact estimate").

5 Hochheiser, *Rohm and Haas*, 97; ROH, 1952 AR, 21–22, 26–28 (28, "question").

6 ROH, 1952 AR, 24, 32 (24, "fits," and 21, "new products").

7 Hochheiser, *Rohm and Haas*, 98; William H. Hill, Sr., and John G. Stauffer, "Rohm and Haas and Latex Paints," *Brush Strokes* 10 (2003), 6–7; Richard E. Harren, "Elements of a Successful Research Project: The Development of an Opaque Polymer," typescript, ca. 1982, 137/6, ROH I.

8 *Chemicals for Industry* (Philadelphia: Rohm and Haas, 1945), 171–172; Charles McBurney, transcript of oral history with Sheldon Hochheiser, 7–10, 11 June 1986, ROH [hereafter cited as McBurney transcript]; "Our Sales Departments: The Resinous Products Division," *Formula* (Apr. 1949), 3, 9.

9 Hochheiser, *Rohm and Haas*, 98; Harren, "Elements"; McBurney transcript, 7 ("Ben Kine").

10 ROH Laboratory 34, "Annual Research Summary," 1951, 4–5, 93/1, ROH; Klein transcript, 7 ("crackerjack"); Robert J. Myers to W. S. Johnson, 8 June 1951; W. C. Prentiss, "Laboratory 23: Report No. 336," 7 May 1951; Myers to Johnson, 30 July 1951; Prentiss, "Laboratory 23: Report No. 347," 9 July 1951, all in 93/1, ROH.

11 Prentiss, "Laboratory 23: Memorandum No. 395," 28 Dec. 1951 ("development"), 93/1, ROH I; "Laboratory 34: Resin Emulsions," 10 April, 6 May, 11 June, and 14 July 1952, 93/1, ROH I.

12 McBurney transcript, 10 ("fellas"); R. W. Auten to Johnson, 24 July ("although") and 18 Sept. 1952, 93/1, ROH I.

13 "Laboratory 34: Resin Emulsions," 20 Aug., 15 Sept., 10 Oct., 14 Nov., and 4 Dec. 1952, and 13 Jan. 1953; Auten to Johnson, 18 Sept. 1952; ROH Laboratory 34, "Annual Research Summary," 1952; Benjamin B. Kine, "Report No. 34-36," 3 June 1953, all in 93/1, ROH I.

14 ROH Laboratory 34, "Annual Research Summary," 1952, 93/1, ROH I.

15 "Rhoplex AC-33 Acrylic Resin Emulsion for Water Paints: Preliminary Notes," C-1-53 and C-2-53, Jan. 1953, 93/1, ROH I.

16 ROH, BM, 22 July 1952, 19–21 ("ruthless"), and 27 Jan. 1953, 17 ("our optimism"); ROH, BM, 21 Apr. 1953, 17 ("sales promotion").

17 Auten to Johnson, 28 Aug. 1953 ("hoping"), 93/1, ROH I; Connor transcript, 35 ("oil paint"); Klein transcript, 9 ("Rhoplex"); ROH, BM, 25 Jan. 1955, 17 ("progress").

18 Carolyn M. Goldstein, *Do It Yourself: Home Improvement in Twentieth Century America* (Washington, DC: National Building Museum, 1998); "Acrylic Paints for 1960," *Reporter* (Jan.-Feb. 1960), 12–13; "The Advantages of Using Acrylic Paint," *Reporter* (July-Aug. 1964), 9–11.

19 Montgomery Ward, *Spring and Summer 1957* (Albany, NY: Montgomery Ward, 1957), 818–819 (quotations); Montgomery Ward, *1963 Spring & Summer* (Denver: Montgomery Ward, 1963), 616–617.

20 "New Paint for Home Decorators," *Reporter* (Jan.-Feb. 1954), 1–6 (5, "advent").

21 "A New Paint for Industry," *Reporter* (Sept.-Oct. 1953), 1–5, 28; "Built on a Gold Mine," *Reporter* (May-June 1956), 2–6; "Perennial Youth for a Pioneer in Paint," *Reporter* (Sept.-Oct. 1958), 6–11.

22 "Producer of Custom Architectural Finishes," *Reporter* (Nov.-Dec. 1964), 11–15 (14, "specialize").

23 ROH, BM, 28 Jan. 1958, 16 ("sales"); 26 July 1960, 11; 24 Jan. 1961, 6; 24 Oct. 1961, 6; 23 Jan. 1962, 7; 17 Apr. 1962, 9; and 21 Apr. 1964, 8; Hochheiser, *Rohm and Haas*, 137–138.

24 Donald C. Garaventi, transcript of oral history with Regina Lee Blaszczyk, 9–10 ("customer"), 14 Nov. 2006, ROH [hereafter cited as Garaventi transcript]; Hill and Stauffer, "Rohm and Haas and Latex Paints," 8; "Superior Paints for Outdoor Wood," *Reporter* (July-Aug. 1963), 14–15; "Rhoplex Acrylic Emulsions Mark 10th Anniversary," *Reporter* (Nov.-Dec. 1963), 3; "New All-Purpose Acrylic Emulsion," *Reporter* (Nov.-Dec. 1964), 3 (AC-22); "The Inside Story of Acrylic Paints," *Reporter* (Jan.-Feb. 1967), 4–8; "A New Rhoplex Polymer for Latex High Gloss Enamels," *Reporter* (Sept.-Oct. 1968), 3; "The Universal Paint Vehicle," *Reporter* (Jan.-Feb. 1969), 3; "Exterior Latex Glass Paints with Long-Lasting Luster," *Reporter* (May-June 1970), 12–15; "Paints with Staying Power," *Reporter* (May-June 1970), 24–25;"Rhoplex AC-34," *Resin Review* 13 (Fall 1962-Winter 1963), 5–17.

25 Garaventi transcript, 10.

26 "Nature's Testing Ground for Paints," *Reporter* (Nov.-Dec. 1961), 18–21; "Test Fences and Test Houses," *Resin Review* 27 (1977), 8; "Happy Birthday, Newtown!," *Resin Review* 28 (1978), 26; "Exposure Station to Celebrate 40th Anniversary," *Brush Strokes* 1 (Spring 1993), 1–5.

27 Shaffer transcript, 29.

28 ROH, BM, 28 Jan. 1974, app., mfm. 7076, Records Management, ROH [hereafter cited as ROH-RM]; Hochheiser, *Rohm and Haas*, 138; Fred W. Shaffer to Vincent L. Gregory,

"Discussion of Our 1977–1979 Business Plan," 4 May 1977 ("North America"), MCM, ROH. Customer Paul Clark from Johnson Paints praised acrylics in "Paints with a Florida Personality," *Reporter* (Spring 1980), 21–23.

29 "The Paint That Conquered Yellow Pine," *Reporter* (May-June 1966), 21–23 (23, "conventional"); Gerould Allyn, "Acrylic Primer for Yellow Southern Pine," *Paint and Varnish Production* (July 1966), 35–39; "Paints with a Florida Personality," *Reporter* (Spring 1980), 23.

30 "The Dutch Boy's Super New Paints," *Reporter* (Spring-Fall 1976), 28–32 (32, "creative"; 29, "exposure").

31 ROH, BM, 28 Jan. 1974, app; "Fine Finishes for the Finns," *Reporter* (Jan.-Feb. 1970), 8–11 (9, "do it yourself"); "Meeting Finland's Paint Needs," *Reporter* (Summer 1985), 6–11.

32 "Changing Spain's Style of Painting," *Reporter* (Winter 1975-76), 8–11 (8, "developments").

33 Gregory transcript (Bohning, 1995), 14 ("advantage"); R. W. Auten, "1971 Annual Report—FECS Tokyo Territory, 72-M-47," 25 Feb. 1972 ("Primal"), 55, box 9-311, ROH-RM.

34 Garaventi transcript, 14.

35 Harren, "Elements"; Garaventi transcript, 14 ("Kowalski").

36 "Management Committee Minutes," 4 June 1980; 28 Jan., 1 June, and 6 Apr. 1981; and 26 Feb. 1982, MCM, ROH; Robert E. Naylor, transcript of oral history with Regina Lee Blaszczyk, 18, Feb. 2007, ROH [hereafter cited as Naylor transcript]; Hill and Stauffer, "Rohm and Haas and Latex Paints," 16; Alexander Kowalski, Martin Vogel, and Robert M. Blankenship (assigned to Rohm and Haas Company), 1982, "Sequential Heteropolymer Dispersion and a Particulate Material Obtainable Therefrom, Useful in Coating Compositions as a Thickening and/or Opacifying Agent," 1982, U.S. Patent 4,427,836, filed 24. Jan. 1982, and issued 24 Jan. 1984; Garaventi transcript, 14–15 ("Ropaque").

37 Hill and Stauffer, "Rohm and Haas and Latex Paints," 13–20; John G. Stauffer and William H. Hill, "Rhoplex Multilobe 200 Emulsions," *Brush Strokes* 1 (Spring 1993), 10–17.

38 Deborah Zimmer, interview by author [hereafter cited as Zimmer interview], 24 July 2008; Al Paul Lefton Company, "Consumer Attitudes toward Exterior House Paint," 4 Dec. 1981, PQI files, ROH; Blaszczyk, *American Consumer Society*, pt. III.

39 Lefton Company, "Rohm and Haas Company: Exterior Architectural Coatings Strategic Marketing Discussion," 9 Mar. 1989; Lefton Company, "Rohm and Haas Quality Acrylic Paint Program," 13 Apr. 1989, PQI Files, ROH.

40 Lefton Company, "Rohm and Haas Company: Exterior Architectural Coatings Strategic Marketing Discussion"("cause"); Acrylic Technology Timeline, ca. 1996, PQI Files, ROH; Zimmer interview, 11 Aug. 2008 ("father").

41 "Quality Paint and Practices," reprint from *American Painting Contractor*, 2, [ca. 1990], PQI files, ROH; Zimmer interview, 11 Aug. 2008; "Research Contributes to Top Quality Paint Performance," *PQI Magazine* 1 (Fall 1992), 12–15, 22–24; "PQI 'Farm' Report," *PQI Magazine* (Winter 1996-97), 12–15.

42 Zimmer interview, 11 Aug. 2008; "Unocal's Polymer Business Sold," *Nonwovens Week* (1 Jan. 1992); ROH, 1992 AR, 8–10; Deborah Zimmer, personal communication with author, 5 Jan. 2009; Donald C. Garaventi, personal communication with author, 31 Dec. 2008.

43 Zimmer interview, 11 Aug. 2008.

44 "PQI Educates and Grows the Home Painting Market," *Formula* (Mar. 2002) (Stauffer quotation); "The Paint Quality Institute: A Brush Stroke of Genius," *Formula* (Oct. 2006).

45 "PQI Educates and Grows the Home Painting Market" (Rogers quotation).

46 "The Paint Quality Institute: A Brush Stroke of Genius" (Applestein quotation)

47 Garaventi transcript, 10.

Rockets at Redstone

On Redstone Arsenal see "Rohm & Haas to Conduct Research for the Army," *Formula* (June 1949), 2 (Haas quote); "Operation Redstone," *Formula* (May 1958), 14–15, 19; "Golden Anniversary 1959," *Formula* (Aug. 1959), n.p. (Connor quote); "Huntsville—Capital of the Space Universe," *Formula* (Jan. 1960), 4–5. On Redstone see Command Historian, U.S. Army Aviation and Missile Command, Redstone Arsenal, "Redstone Arsenal Complex Chronology," on Redstone Historical Information Web site, http://www.redstone.army.mil/history/install/install.html <accessed 27 Dec. 2008>.

Down Under

Primal Chemicals, *Annual Report 1964*, Table 2 (1963–1964 sales), Resins section, n.p. ("Lo-Gloss"), box 11-987, ROH-RM; Primal Chemicals, *Annual Report 01966*, Coatings section, 1–7 (1, "twelve"), in box 9-340, ROH-RM; Rohm and Haas Australia, *Annual Report 1973*, 26 (sales figures), 35 (quotations), in box 67H-02, ROH-RM; "Fine Paints from 'Down Under,' " *Reporter* (Jan.-Feb. 1966), 21–25.

Chapter 7

1 Gregory transcript (Ozer, 1988), 21.

2 "Rohm and Haas's Post-Op Growth Plan," *Chemical Week* (27 Jan. 1982), 46–50.

3 Gregory, "Management Committee Meeting—Minutes, September 14, 1976," MCM, ROH; Shaffer and A. J. Ballard, "Management Committee Meeting, February 1, 1977," MCM, ROH (Shaffer quotation).

4 "SCI Medalist Gregory Discusses Industry Experiences and Concerns," *Chemical and Engineering News* (10 Oct. 1988), 9–12 (9, "expansion"); ROH, 1970 AR, 1, and 1971 AR, 1; Gregory transcript (Ozer, 1988), 16, 21–22; Gregory transcript (Bohning, 1995), 31.

5 Geoffrey Smith, "A Funny Thing Happened," *Forbes* (19 Mar. 1979), 64, 66 (66, "glowing talk"); Gregory transcript (Ozer, 1988), 19; Gregory, "Company Report," 1 ("illusions"), in ROH, *Sixty-Second Annual Meeting of Stockholders*, 5 May 1980, ROH.

6 Gregory transcript (Ozer, 1988), 21; Gregory transcript (Bohning, 1995), 30–31; C. J. Prizer, "Organizational Group Study Report," 1 May 1975, 9/1, ROH.

7 Shaffer transcript, 19.

8 Shaffer, "Discussion of Rohm and Haas Business Plan for 1977–79, Meeting of May 2, 1977," MCM, ROH; Wilson transcript (2006), 30.

9 David Gass, market analyst's report on Rohm and Haas for Drexel Burnham Lambert, New York, 30 Sept. 1980, 20, ROH.

10 Shaffer, "Discussion of Rohm and Haas Business Plan."

11 "Kulik Says Goodbye to Rohm and Haas," *Formula* (July 1993), 2, 4.

12 Gregory transcript (Ozer, 1988), 16–20; Hochheiser, *Rohm and Haas*, 164, 168–183, 191–192.

13 The New Ventures Business Team was proposed in 1976 and created in 1978. See Gregory, "Management Committee Meeting—July 26–27, 1976"; Shaffer, "Management Committee Minutes—September 9, 1977"; and Shaffer, "Management Committee Minutes—September 30 and October 3, 1977" ("heavy responsibility"), MCM, ROH.

14 Shaffer, "Management Committee Minutes—July 27, 1979" ("new investment"), MCM, ROH.

15 Shaffer, "Management Committee Meeting—Long Range Plan 1980–1989—Sept. 17 & 18, 1979" ("principal"), MCM, ROH.

16 Wilson transcript (2006), 31 ("knitting").

17 Shaffer, "Management Committee Meeting—October 26, 1979" ("57 sectors"); Shaffer, "Management Committee Minutes—September 26, 1981," MCM, ROH.

18 "Firms Take Short-Cut to New Technology," *Chemical Week* (2 Jan. 1980), 33; "Rapid Wrap-Up," *Chemical Week* (21 May 1980), clipping, Gregory bio file, ROH.

19 "Great Chemistry: Richard Shipley," *The Manager* (Fall 1999), 17–20; Joan Rousseau, interview by author, 22 Aug. 2008.

20 Deborah W. Hairston, "The Shipley Partners: Good Chemistry at Work," *Chemical Week* (22 May 1985), 26–27; "Lucia Farrington Wed," *New York Times*, 12 Oct. 1941; "Electronics Used to Speed Invoices," *New York Times*, 3 Mar. 1955; "Computers Get Aid That Counts," *New York Times*, 14 June 1959; "Charles R. Shipley Jr., 86, Innovator in Electronics," *Boston Globe*, 22 June 2004; George Reed, *The Story of the Shipley Company, 1957–1992* ([Newton, MA: Shipley Company, 1992)], 2–3, ROH.

21 Reed, *Story of Shipley Company*, 4; Charles R. Shipley, Jr., transcript of interview with George Reed, 25 Aug. 1988 [hereafter cited as Charles R. Shipley, Jr., transcript], 2 ("entire concept"), Shipley Archives, Rohm and Haas Electronic Materials, Marlborough, Mass. [hereafter cited as SA]; Lucia Shipley, "The Early Years," *Shipley Newsletter* (Summer 1992), 1; Robert L. Goldberg, "Submission in Support of Nomination of Charles R. Shipley Jr. for an Honorary Doctoral Degree," 2, 8 Dec. 1982, SA.

22 Lucia H. Shipley, transcript of interview with George Reed, 3, 23 Aug. 1988, SA.

23 Hairston, "The Shipley Partners" ("decision"); Reed, *Story of Shipley Company*, 8.

24 Charles R. Shipley, Jr., transcript, 6 ("Remington Rand"); Reed, *Story of Shipley Company*, 8–10.

25 Charles R. Shipley, Jr., transcript, 6 ("kiss").

26 Goldberg, "Submission," 2–3; George Reed and Carmen DeSisto, "Paradigm Shifts at Shipley," *Quality at Shipley* (Aug.-Sept. 1991), 5.

27 David B. Arnold, transcript of interview with George Reed, 6, 8 Sept. 1988, SA [hereafter cited as Arnold transcript]; Reed, *Story of Shipley Company*, 11, 13; Goldberg, "Submission," 4–6.

28 Howard A. MacKay, "Electroless Nickel Plating," *Shipley Newsletter* (Aug. 1985), 1–2. Plastic PC housings were coated with electroless copper and nickel to prevent interference with other electronic devices, such as televisions and pacemakers. See Reed, *Story of Shipley Company*, 47; Mark Farsi, "EMI Success," *Shipley Newsletter* (Mar. 1987), 1; Reed, "How Shipley Products Are Used in the Industry," *Shipley Newsletter* (Autumn 1991), 3; Farsi, "A One-Sided Solution to EMI Shielding," *Shipley Catalyst* (Feb. 1990), 3; Russell House, "The Art of Deception," *Shipley Catalyst* (Autumn 1992), 4.

29 Reed, *Story of Shipley Company*, 18–19, A10–A11; Goldberg, "Submission," 1–2, 6–7; Charles R. Shipley, Jr., "Historical Highlights of Electroless Plating," *Plating and Surface Finishing* (June 1984), offprint, SA.

30 David B. Arnold, "Shipley Company—Foreign History," 8 Oct. 1969, 5 (quotation), SA; Reed, *Story of Shipley Company*, A2–A7.

31 Reed, *Story of Shipley Company*, 38–41; Richard C. Shipley, transcript of oral history with Regina Lee Blaszczyk and Arnold Thackray, 7, 2 Nov. 2006, ROH [hereafter cited as Richard C. Shipley transcript].

32 Richard C. Shipley transcript, 8–11 (quotation); Arnold transcript, 11.

33 Arnold transcript, 11–12.

34 PR Newswire, 8 Jan. 1982, LexisNexis Academic (Gregory quotation); "Briefs," *New York Times*, 9 Jan. 1982; "Rohm & Haas Agrees to Acquire 30% Stake in Chemicals Concern," *Wall Street Journal*, 11 Jan. 1982; "Rohm and Haas Acquires 30% Minority Interest in Shipley," *Shipley Newsletter* (Summer 1982), 3; ROH, BM, 7 Feb. 1983, 6, mfm. 19599, ROH-RM; ROH, 1982 AR, 2–3 ("splurge").

35 Wilson, "Talks to Groups Arranged by Salomon Brothers," July 1983, 1, Wilson bio file, ROH; "Chemicals," *Forbes* (13 Jan. 1986), 83–84 (83, "plain fact").

36 "An Interview with Vincent L. Gregory, Jr.," *HBS Bulletin* (Dec. 1988), 50–55; ROH, BM, 22 Sept. 1983, 2–4 ("emphasis on efficiency"), mfm. 19599, ROH-RM.

37 "An Interview with Vincent L. Gregory, Jr.," 52 ("competition"); "Ford Embraces Six-Sigma Quality Goals," Society of Manufacturing Engineers Newsdesk, 13 June 2001, http://www.sme.org/cgi-bin/get-press.pl?&&20012513&ND&&SME& <accessed 10 Nov. 2008>. For his take on Deming see Gregory, "Prior Year Results," in ROH, *Company Report: 70th Annual Meeting of Stockholders*, 2 May 1988, 6, ROH.

38 "Special Report: Quality," *Formula* (May 1985), 3–4 (3, Djuvik quotation).

39 "One Short Move for EMCA," *Reporter* (Fall 1989), 32–33; "The Road Ahead for Rohm and Haas," *Reporter* (Spring 1985), 9–13; Pamela T. Richards, "Shipley & Harris Semiconductor SPC Partnership," *Shipley Newsletter* (Aug. 1987), 2; L. Shipley, "Shipley Quality," *Quality at Shipley (QS)* (Jan.-Feb. 1992), 1; "Quality at Shipley and ISO 9000," *Shipley Newsletter* (Autumn 1993), 1; quotation in L. Shipley, "Quality," 13 July 1982, SA; Richard C. Shipley transcript, 23; *Shipley* (Newton, MA: Shipley, ca. 1987), 1; Bob Marckini, "What Is the Quality Process?" *QS* (Feb. 1986), 1–2; L. Shipley, "New Name Contest," *QS* (Sept. 1986), 1; Charlie Shipley, "A New Era at Shipley," *QS* (Apr. 1986), 1; Dick Shipley, "Constancy of Purpose," *QS* (May 1986), 1; L. Shipley, "Progress Continues," *QS* (June 1986), 1; R. C. Shipley, "A Year in Review," *QS* (Dec. 1986), 1; Dick Shipley, "Corporate CIP Objective," *QS* (Jan. 1988), 1; Richard C. Shipley to author, 12 Nov. 2008; "TCT Time Cycle," *Shipley Newsletter* (Autumn 1992), 1.

40 "An Interview with Vincent L. Gregory," 52–53 (52, quotations); J. P. Mulroney, "Operations Report," 9, in ROH, *Company Report: 69th Annual Meeting of Stockholders*, 4 May 1987, ROH.

41 Arnold, "Shipley Company—Foreign History," 10–11; Reed, "Shipley Far East Ltd.: A Fifteen-Year History of Success," *Shipley Newsletter* (Summer 1991), 1–2; Richard C. Shipley transcript, 13 ("Japanese subsidiary").

42 Steve Hudson, "World-Class Competitor," *Shipley Newsletter* (Apr. 1986), 4.

43 Jim Murphy, "How to Cut Business Travel Costs at Shipley," *Quality at Shipley* (Nov. 1988), 2; "New Shipley Facility," *Shipley Newsletter (SN)* (Apr. 1986), 8; "Shipley Opens Sales Office in Hong Kong," *SN* (Aug. 1987), 1; "Technical Training Programs in Southeast Asia," *SN* (Nov. 1987), 3; "Shipley Chemicals Pte. Ltd. Opening Party," *SN* (Aug. 1985), 2; "Sasagami Phase II Completed," *SN* (Nov. 1986), 1; George Reed, "Globalization—A State of Mind," *Quality at Shipley* (Jan.-Feb. 1992), 2 (Reed quotations).

44 On the Wilson-Mulroney team see "Adding Value to Rohm and Haas," *Chemical Week* (7 June 1989), 41–42.

45 Wilson transcript (2006), 34.

46 Ibid., 34, 37.

47 Wilson transcript (2006), 31, 38–40 (38, "defensible"); "Rohm and Haas Sees Earnings Bloom, but Returns Disappoint," *Chemical Week* (21 Sept. 1994), 42, 44; Richard C. Shipley transcript (2006), 15; Naylor transcript, 19.

48 Rajiv L. Gupta, transcript of oral history with Regina Lee Blaszczyk and Arnold Thackray, 3 Nov. 2006 and 7 Jan. 2007, 50, ROH [hereafter cited as Gupta transcript].

Blazer

"Gregory Says 'Exciting Things' Could Lead to Golden Era," *Formula* (April 1981), 1 ("more"); "Protecting Number One," *Reporter* (Autumn 1980), 18–23; "Discovering a Herbicide," *Reporter* (Winter 1980-81), 20–23 (20, Unger quote); "Getting Blazer to 'Go,' " *Reporter* (Spring 1981), 2–5; L. Upton Hatch, "Innovation and Regulation in the Pesticide Industry: Four Case Studies," P83-14, Staff Paper Series, Department of Agricultural and Applied Economics, Institute of Agriculture, Forestry and Home Economics, University of Minnesota, St. Paul, MN, June 1983; "Right on Target," *Reporter* (Summer 1986), 16–19 (18–19, Myers); Gregory transcript (Bohning, 1995), 32–33 (33, "success").

Next Gen

Ellen Goldbaum, "Rohm & Haas: Adding Value to a Proven Formula," *Chemical Week* (7 June 1989), 41–43 (41, "$1 billion"); Marc S. Reisch, "SCI Medal Goes to J. Lawrence Wilson," *Chemical & Engineering News* (11 Oct. 1999), 31–36 (35, "solve problems"); Wilson transcript (2006) and (1999).

Chapter 8

1 "A Successful Joint Venture," *Reporter* (Spring 1973), 26–29; "India to Consider Rohm and Haas's Proposed Re-Entry," *ICIS News* (28 Nov. 1994); Edward L. Stanley to J. R. Traverner, "Indonesia Project: Tentative Conclusions," 11 July 1972, ROH. Some early Asian ventures were short-lived. A Chinese sales office established through the Taipei branch was closed when the Taiwanese government objected to the U.S. agent representing two companies. See ROH, BM, 7 Feb. 1975, 13–14, mfm. 7076, ROH-RM.

2 "Poland: A New Industrial Power," *Reporter* (Spring 1977), 15–19.

3 "A Multifaceted Multinational Company," *Reporter* (Spring 1973), 4–8 (8, Felley quotation); "Joint Venture in Yugoslavia," *Reporter* (Fall 1975), 16–21 (20, Jub quotation); "Rohm and Haas Company," in *The World Directory of Multinational Enterprises*, John M. Stopford, John H. Dunning, and Klaus O. Haberich (New York: Facts on File, 1980), 871–873.

4 "Agreements Signed with Soviet Union," *Reporter* (Summer 1975), 4–7 (6, "adapt and formulate").

5 Wilson transcript (1999), 13 ("worldwide"); ROH, BM, 27 Jan. 1975, 13–15, Schedules I-II, mfm. 7076, ROH-RM; Hochheiser, *Rohm and Haas*, 149–150.

6 Wilson transcript (1999), 13 ("one-world thinking").

7 Shaffer, "Management Committee Minutes—October 27, 1978," MCM, ROH.

8 Shaffer, "Minutes—Management Committee Meeting—October 2–4, 1980," MCM, ROH.

9 Michael McCoy, "Rohm and Haas Picks Two for Top Posts," *Chemical & Engineering News* (3 Aug. 1998), 8; "Rohm and Haas: Top Three in Specialties," *Chemical Week* (17 Feb. 1999), 29–32. On the careers of Fitzpatrick and his wife, Jean, see J. Michael Fitzpatrick, transcript of oral history with Regina Lee Blaszczyk, 15 Dec. 2006, ROH [hereafter cited as Fitzpatrick transcript].

10 Michael McCoy, "Plans Pay Off for Rohm and Haas," *Chemical & Engineering News* (22 Feb. 1999), 13–15.

11 Prasad Gune, "Raj Gupta at Rohm and Haas," research paper for "The Coming of Managerial Capitalism," Harvard Business School, Dec. 1999, Gupta bio file, ROH.

12 Gune, "Raj Gupta at Rohm and Haas"; Gupta transcript, 29–32; Vassiliou transcript, 8–9; Riddhi Trivedi-St. Clair, "An Outstanding South Asian Success Story," *Biz India* (Mar.-Apr. 2006), 10–11, 27 (11, "French"). On Duolite see, for example, ROH, BM, 13 June 1983, 1–2; 25 July 1983, 4–6; and 23 Apr. 1984, 8–10, all on mfm. 19599, ROH-RM.

13 Allen Levantin, "European Region," in ROH, BM, 18 June 1984, 8–9, mfm. 19599, ROH-RM; Gupta transcript, 29 ("efficient").

14 Gupta transcript, 32 ("global organization"); Wilson transcript (2006), 45.

15 Gupta transcript, 34.

16 Gupta transcript, 34–35.

17 Gupta transcript, 36–37; Wilson transcript (2006), 34; Richard G. Barnes, "The Core Beliefs of a Global Leader," *Lead Magazine* (Spring-Summer 2007), 8–13; Gupta transcript, 36–37.

18 Wilson transcript (1999), 9, and (2006), 20; Marc S. Reisch, "SCI Medal Goes to J. Lawrence Wilson," *Chemical & Engineering News* (11 Oct. 1999), 31–36; Gupta transcript, 40–41.

19 Andrew Wood, "Rohm and Haas Expands Acrylics Business in China," *Chemical Week* (31 Aug.-7 Sept. 1994), 19; David Rothman and Andrew Wood, "Firms Look to Tap China's Research," *Chemical Week* (31 Aug.-7 Sept. 1994), 26; Gupta transcript, 40–41; Wilfried Vanhonacker, "Entering China: An Unconventional Approach," *Harvard Business Review* (Mar.-Apr. 1997), 130–140; C. K. Prahalad and Kenneth Lieberthal, "The End of Corporate Imperialism," *Harvard Business Review* (July-Aug. 1998), 68–79.

20 Gupta transcript, iii, 37–38.

21 Gupta transcript, 37–41; Alan Tyler, "Asia Seeks Slice of American Pie," *Asia-Pacific Chemicals* (Oct. 1995), 32–36.

22 Gupta transcript, 40–41, and 37, for his description of Wilson's foresight about Asia.

23 Wilson transcript (1999), 13–14 ("operating management"); Vassiliou transcript, iii; Garaventi transcript, iii.

24 Shaffer transcript, 22; Rajiv L. Gupta, "Asia-Pacific," *Polymer News* 21 (1996), 402–404 (403, Grehl quotation).

25 Gupta transcript, 37 ("instinct"); Rajiv L. Gupta to Corporate Leadership Council, "Asia-Pacific: An Opportunity and a Challenge," 10 Jan. 1996, Executive Files, ROH.

26 On electronics see "Plans Pay Off for Rohm and Haas," *Chemical & Engineering News* (22 Feb. 1999), 13–15; Dave Schram, telephone interview with author, 13 Nov. 2008; "Company News: Rohm & Haas to Buy LeaRonal for $460 Million," *New York Times*, 22 Dec. 1998; "LeaRonal," *International Directory of Company Histories*, vol. 23 (Chicago: St. James Press, 1998); Rajiv L. Gupta, remarks to the Schroders/ *Chemical Week* Chemical Industry Conference, 8 Dec. 1998, Gupta bio file, ROH; Rajiv L. Gupta, "Rohm and Haas Company," presentation to Goldman, Sachs and Co., Sixth Annual Chemical Investors Forum, Key Largo, Florida, 22 May 1998, Gupta bio file, ROH; Wilson transcript (2006), 38–40, 46.

27 Briggs-Gammon, "Rohm & Haas Shifts to the Future"; Andy Gotlieb, "Refocused Rohm Cooks Up Growth," *Philadelphia Business Journal* 21 (5-11 Apr. 2002), 1, 40.

28 "Pierre R. Brondeau," Dec. 2008, ROH.

29 "Interview with Raj L. Gupta: India Has an Edge in Knowledge-based Industries," *Business Today* (26 Oct. 2003), 80–83; quotation in "One Step Beyond," *ECN: European Chemical News* (4 Aug. 2003), clipping, Gupta bio file, ROH.

30 Harold Brubaker, "Restructuring Is on Course, CEO of Rohm & Haas Says," *Philadelphia Inquirer*, 9 Oct. 2001, D5; "Playing to New Rules," *Chemical Week* (14 May 2003), 5.

31 Kerri Walsh, "Making Specialty Chemicals Special Again," *Chemical Week* (26 June 2002), 19–23; Cynthia Challenger, "Executive Insight: Rohm and Haas Stresses R&D, Cost-Cutting in Managing for Recovery and Growth," *Chemical Market Reporter* (29 Apr. 2002), sec. 3, 21, clipping, Gupta bio file, ROH; Andrew Wood and David Hunter, "Rohm and Haas: Working on the Margins," *Chemical Week* (30 Apr. 2003), 18–20; Fitzpatrick transcript, 21–22; Chana R. Schoenberger, "Saved by Software," *Forbes* (23 Dec. 2003), clipping, Fitzpatrick bio file, ROH.

32 Rajiv L. Gupta, "Growth through Innovation: The Specialty Chemical Perspective," *Polymer News* 28 (2003), 170–173 (172, "technology").

33 Ibid.

34 For Gupta, on the need to move away from standardized products in the global environment, see "Vision 2010: Positioning for Continued Success," *Formula* (Nov. 2006), 1; "Different Strokes," *ASIAN Chemical News* (28 Apr.-4 May 2003), I–II; "Poised for Bigger Growth," *ASIAN Chemical News* (28 Apr.-4 May 2003), III–IV (IV, "consultant").

35 Robert Westervelt, "Rohm and Haas: Capturing Emerging Opportunities," *Chemical Week* (25 Oct. 2006), 21–24.

36 Ibid, 21.

37 J. Robert Warren, "A Look Offshore," *Chemical Marketing Reporter* (24 July 1995), clipping, Gupta bio file, ROH; "Rohm and Haas: Capturing Emerging Opportunities," 21.

38 "Vision 2010."

Yankees Abroad

Fitzpatrick interview, 9–14 (10, "virtually"; 11, "lots," "*selling*"; 12, "environments"; 13–14, "hyperinflation").

Raj Gupta: Numbers Guy

Gupta transcript, 26 ("idea"), 28 ("U.K. citizens"), 32 (stores), 35 ("people").

Index

Boldface indicates an image.

About the Author

Regina Lee Blaszczyk, Ph.D., is an award-winning corporate historian who specializes in the history of innovation. She is the author or editor of six books, including *Imagining Consumers: Design and Innovation from Wedgwood to Corning*; *Major Problems in American Business History: Documents and Essays*; *Producing Fashion: Commerce, Culture, and Consumers*; and *American Consumer Society, 1865–2005: From Hearth to HDTV*. In 2008 she received the Harold F. Williamson Prize, a mid-career award from the Business History Conference, the largest international organization of business historians. Visit her Web site: www.imaginingconsumers.com.

Photography Credits

All imagery is from the collections of Rohm and Haas Company and its employees and alumni, and is used by permission of Rohm and Haas, except for:

American Sign Museum, Cincinnati, Ohio: 116
Regina Lee Blaszczyk: 26 (object), 34 (object), 133, 135, and 138
Keith Brofsky: 8, 136–137, 152–153, 158, 160–161, and 168–169
Ian Campbell Photography: 5, 20–21, 28–29, 44–45, 52–53, 66–67, 72–73, 86–87, 94–95, 112–113, 122–123, 144–145, 182–183, 191, 192–193, and 197
Courtesy of the Chemical Heritage Foundation Collections: 55
Cooper-Hewitt, National Design Museum, Smithsonian Institution, Gilbert Rohde Collection. Gift of Lee M. Rohde. Copy photo: Matt Flynn: 60, 62, and 64
Courtesy of Evonik Industries AG, Corporate Archives: 17, 19 (object), 50, 65, and 117
General Motors Corp. Used with permission, GM Media Archives: 63 (photograph) and 108
Courtesy of Hagley Museum and Library: 16 (object), 27, and 34 (photograph)
O. Louis Mazzatenta/Getty Images: 147
Museum of History and Industry, Seattle, Washington: 124
Charlie Simokaitis Photography: 189
Courtesy of Smithsonian Institution Libraries, Washington, DC: 68
The W. Edwards Deming Institute: 173

All photographs of archival prints, documents, and objects were taken by Fenwick Publishing unless otherwise noted.

1909-2009
YEARS
A history of innovation. A lifetime of inspiration.

ROHM AND HAAS